AF801652

Crystal Research and Technology

4

Journal of Experimental and Industrial Crystallography

Zeitschrift für experimentelle und technische Kristallographie

Volume 23
1988
Number 4 April

Akademie-Verlag · Berlin

ISSN 0232-1300 Cryst. Res. Technol., Berlin 23 (1988) 4, 465–582, K63–K79

Directions to the author

In the journal "Crystal Research and Technology — Kristall und Technik" as a rule the following categorie of papers will be printed: original papers, review articles, short notes, and book reviews, **preferentially in English.** Complete information necessary for type-setting and a "Short Title" should be added.

The manuscript should not exceed 10 pages and be type-written on one side only in the format A 4 (21 × 29,7 cm) on white ink-proof paper, with a two-lined distance and a left-hand free edge of 3 cm, and submitted in 2 copies. All sheets of the paper belonging to the contribution are continuously to be numbered.

The head of the manuscript should be arranged in this manner: initials and name of the author; institution, and title of the paper. Then two abstracts should follow, one in the original language and the second in German, English, French, or Russian, respectively.

Formulae, signs and symbols, letters etc. being wanting on the type-writer, should be distinctly written with penholder. O as zero is to be underlined, the capital O is not. Greek letters should be underlined with red ink.

Paragraphs should be marked by indenting of at least 3 touches.

Concerning the marking of the manuscript should be noted:
headings should be underlined, sub-headings not, and indented by 9 touches. The titles of tables are not to be underlined. Other markings in the text will be done by the editorial staff. They mean:

light-red ∼∼∼∼∼ = small capitals	light-red	perpendicular straight red line on the left edge = small printing
blue ∼∼∼∼∼ = in italics		

Tables are to be added on separate sheets of paper at the end of the contribution.

Foot notes should be continuously written on a separate sheet of paper.

Figures (Fig.) — drawings and photographs — should be not more than 8 and submitted (2 copies) in a special envelope: photographs as white high-gloss prints (inscription on a transparent covering sheet) and drawings as distinct sketches. On the back of the submitted figures the reduction scale in tenth (9/10, 8/10, 5/10 e.g.), author's name, short title of the paper, and the number of the figures should be given.

Captions for the figures should be submitted on a separate sheet of paper.

References should be written without numbers in alphabetical order of the names at the end of the contribution. Cited names in the text stand for the references. The references should contain the following items, for example:

for books: RAST, K.: Physikalisch-chemische Rechnungen, Berlin 1958
for journals: KRAUSE, H. N., MÜLLER, K.: Cryst. Res. Technol. **16** (1981) 70

Add, please, the short title, running title, about 6 keywords, and the academic grades of the authors.
The authors will be sent two galley proofs of review articles and original papers. Proof reading of short notes and book reviews will be done by the editorial staff.

For each of these papers a total of 50 reprints is available free of charge. Requests for additional maximum 200 reprints (against payment) should be given together with the submitted manuscript.

Founded 1966 as
Kristall und Technik

Crystal Research and Technology 4

VOLUME 23 · 1988 · NUMBER 4

Journal of Experimental and Industrial Crystallography

ESTABLISHED BY
W. KLEBER and H. NEELS

EDITED BY
H. NEELS, Leipzig

EDITORIAL BOARD
J. BOHM, Berlin; J. HEYDENREICH, Halle
P. PAUFLER, Leipzig; W. SCHATT, Dresden
M. SCHENK, Berlin

ADVISORY BOARD
L. N. ALEKSANDROV, Novosibirsk
Č. BARTA, Prague; F. BERTAUT, Grenoble
G. BLIZNAKOV, Sofia
O. BRÜMMER, Halle
V. A. FRANK-KAMENETZKIJ, Leningrad
S. HAUSSÜHL, Köln
R. KAISCHEW, Sofia; R. KERN, Marseille
Y. KOMURA, Hiroshima; I. KOSTOV, Sofia
P. KRISHNA, Varanasi
K. ŁUKASZEWICZ, Wrocław; K. MEYER, Berlin
I. S. REZ, Moscow
D. SCHULZE, Dresden; N. N. SIROTA, Moscow
I. TARJAN, Budapest
O. M. TODES, Leningrad
A. A. TSCHERNOV, Moscow
A. F. WITT, Cambridge (U.S.A.)
W. A. WOOSTER, Cambridge

EDITOR-IN-CHIEF
H. NEELS, Leipzig

EDITORIAL OFFICE
A. DREILICH
R. NEELS
B. SCHUMANN

THEORIA CUM PRAXI

Akademie-Verlag · Berlin

Papers should be directed to the editor's office
Scharnhorststraße 20, DDR-7030 Leipzig,
or to the Editor-in-Chief Prof. Dr. Hermann Neels,
Zeititzer Weg 25, DDR-7254 Machern,
or to one of the members of the Editorial Board or Advisory Board.

Orders can be sent

- in the GDR:
  to the Postzeitungsvertrieb or to the Akademie-Verlag Berlin, Leipziger Straße 3 – 4, DDR-1086 Berlin;
- in the other socialist countries: to a book-shop for foreign language literature or to the competent news-distributing agency;
- in the FRG and Berlin (West): to a book-shop or to the wholesale distributing agency Kunst und Wissen, Erich Bieber OHG, Wilhelmstr. 4 – 6, D-7000 Stuttgart 1;
- in the other Western European countries: to Kunst und Wissen, Erich Bieber GmbH, General Wille-Str. 4, CH-8002 Zürich;
- in USA and Canada: to the VCH Publishers, Inc., 303 NW 12th Avenue, Deerfield Beach, FL 33442-1788, USA;
- in other countries; to the international book- and journal-selling trade, to Buchexport, Volkseigener Außenhandelsbetrieb der DDR, Postfach 160, DDR-7010 Leipzig, or to the Akademie-Verlag Berlin, Leipziger Straße 3 – 4, DDR-1086 Berlin.

Journal „Crystal Research and Technology“

Herausgeber: Im Auftrage des Akademie-Verlages Berlin herausgegeben von Prof. Dr. Hermann Neels.

Verlag: Akademie-Verlag Berlin, Leipziger Str. 3 – 4, DDR-1086 Berlin; Fernruf: 2236221 oder 2236229; Telex-Nr.: 114420; Bank: Staatsbank der DDR, Berlin, Kto.-Nr. 6836-26-20712.

Chefredakteur: Prof. Dr. Hermann Neels, Zeititzer Weg 25, DDR-7254 Machern; Fernruf: Brandis 3442.
Anschrift der Redaktion: Scharnhorststr. 20, DDR-7030 Leipzig; Telefon: 310502.

Veröffentlicht unter der Lizenznummer 1526 des Presseamtes beim Vorsitzenden des Ministerrates der Deutschen Demokratischen Republik.

Gesamtherstellung: VEB Druckerei „Thomas Müntzer“, DDR-5820 Bad Langensalza.

Erscheinungsweise: Das Journal „Crystal Research and Technology“ erscheint monatlich. Die 12 Hefte eines Jahrganges bilden einen Band. Bezugspreis je Band 480,– DM zuzüglich Versandspesen; Preis je Heft 40,– DM. Der gültige Jahresbezugspreis für die DDR ist der Postzeitungsliste zu entnehmen.

Bestellnummer dieses Heftes: 1082/23/4.

Urheberrechte: Alle Rechte vorbehalten, insbesondere die der Übersetzung. Kein Teil dieser Zeitschrift darf in irgendeiner Form – durch Photokopie, Mikrofilm oder irgend ein anderes Verfahren – ohne schriftliche Genehmigung des Verlages reproduziert oder in eine von Maschinen, insbesondere von Datenverarbeitungsanlagen verwendbare Sprache übertragen oder übersetzt werden. – All rights reserved (including those of translations into foreign languages). No part of this issue may be reproduced in any form, by photoprint, microfilm, or any other means, nor transmitted or translated into a machine language, without written permission from the publishers.
© 1988 by Akademie-Verlag Berlin. Printed in the German Democratic Republic.
AN (EDV) 35830
02800

Cryst. Res. Technol.	23	1988	4	467–474

Original Papers

P. REICHE, J. BOHM, B. HERMONEIT, P. RUDOLPH, D. SCHULTZE

Zentralinstitut für Optik und Spektroskopie,
Akademie der Wissenschaften der DDR, Berlin-Adlershof, DDR, and
Sektion Physik, Humboldt-Universität, Berlin, DDR

Wachstumsmorphologie von $LiNbO_3$-Einkristallen bei der Züchtung nach dem Czochralski-Verfahren unter der Einwirkung eines elektrischen Feldes

Herrn Professor Dr. Hermann NEELS zum 75. Geburtstag gewidmet

Die Wachstumsmorphologie von Lithiumniobat-Einkristallen, die nach der Czochralski-Methode gezüchtet wurden, wird als Ergebnis des Wechselspiels von atomar rauhen und atomar glatten (Facetten-)Flächen beschrieben. An glatten Flächen ist die Grenzflächenspannung zwischen Kristall und Schmelze merklich größer als an rauhen Flächen, was bei bestimmten Orientierungen der Facetten zu einer kapillaren Depression des Meniskus führen kann. Daraus resultierende hydrostatische Instabilitäten werden als Ursache für die spontane Keimbildung und Entstehung von Polykristallen angesehen.

Durch Anlegen eines elektrischen Feldes zwischen Kristall und Schmelze wird die Grenzflächenspannung verringert. Das führt unter anderem zu verbesserter Benetzung der Facetten durch die Schmelze, die Entstehung der Polykristalle wird unterdrückt.

The growth morphology of Czochralski-grown lithium niobate single crystals is described as resulting from interactions of atomary rough and atomary smooth surfaces (facets). The interfacial tension between crystal and melt at smooth faces is considerably larger than at rough ones, causing for appropriately oriented facets a capillary depression of the meniscus. Due to the resulting hydrostatic instabilities spontaneous nucleation may occur that leads to the formation of polycrystals. If an electrical field is applied between crystal and melt the interface tension is reduced. Among other effects, the wetting of facets by the melt is improved and the formation of polycrystals is suppressed.

Einleitung

Die Einkristallzüchtung von Lithiumniobat $LiNbO_3$ nach dem Czochralski-Verfahren ist schon wiederholt und auch zusammenfassend beschrieben worden (RÄUBER). Ein morphologisches Charakteristikum der gezüchteten Kristalle sind Kämme (engl. ridges) entlang ihrer Oberfläche. Sie entstehen als Folge von Facetten, welche sich am Rand der Wachstumsfront ausbilden (NIIZEKI et al.). Die Verteilung der Kämme folgt der Symmetrie des Kristalls bezüglich der Ziehrichtung; ihr ungestörter Verlauf ist ein Indiz, daß der Kristall monokristallin wächst. Die Züchtungspraxis hat sich mit dem Problem auseinanderzusetzen, daß es zuweilen spontan zu einem polykristallinen Wachstum kommt, das bevorzugt im Bereich der Kämme einsetzt. Besonders kritisch ist die Phase der Verbreiterung des Kristalls zu Beginn des Züchtungsvorganges.

Wachstumsmorphologie von *y*-Kristallen

Das ferroelektrische Lithiumniobat gehört bei Zimmertemperatur zur (polaren) Kristallklasse $3m$. Die Wachstumsmorphologie entspricht jedoch der Symmetrie der paraelektrischen Hochtemperaturphase, d. h. der (zentrosymmetrischen) Kristallklasse $\bar{3}m$. Für die Züchtungspraxis von Belang sind (u. a.) Kristalle, die in Richtung der y-Achse gezogen werden („y-Kristalle"). Die y-Achse steht senkrecht zur kristallographischen c-Achse (z-Achse) und zur kristallographischen a-Achse (x-Achse) und entspricht der Richtung [12.0] senkrecht auf $(01\bar{1}0)$ in hexagonaler (Bravaisscher) Indizierung. In Übereinstimmung mit der Spiegelsymmetrie dieser Ziehrichtung (Symmetriegruppe m) wird der Querschnitt des Kristalls nicht genau kreisrund, sondern schwach elliptisch mit dem größeren Durchmesser in der Richtung der x-Achse bzw. [10.0], und auch die Verteilung der Kämme ist spiegelsymmetrisch (Fig. 1).

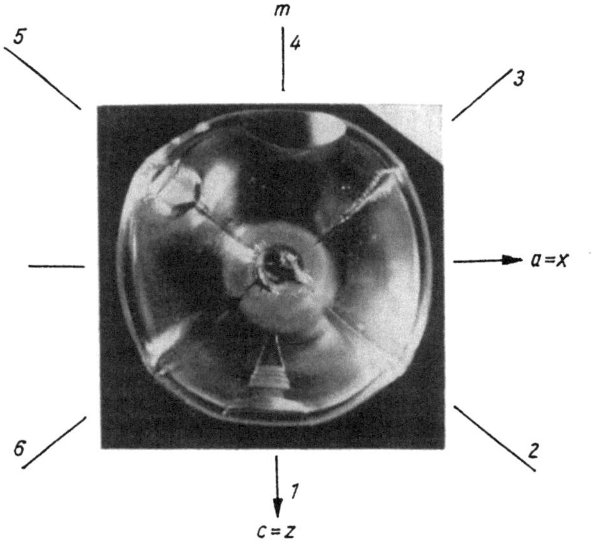

Fig. 1. Kopfbild eines $LiNbO_3$-Kristalls, gezogen in Richtung [12.0] (y-Achse), Maßstab 1:1,5. 1 Facetten $(01\bar{1}2)$ (oben) u. (0001), Zone [10.0], 2 Facette $(1\bar{1}02)$ (unten), Zone $[20.\bar{1}]$, 3 Facette $(10\bar{1}\bar{2})$ (oben), Zone [20.1], 4 Facetten $(0\bar{1}1\bar{2})$ (unten) u. $(000\bar{1})$, Zone [10.0], 5 Facette $(\bar{1}10\bar{2})$ (oben), Zone $[20.\bar{1}]$, 6 Facette $(\bar{1}012)$ (unten), Zone [20.1]

Kämme und Facetten sind eine auch von anderen Kristallen (Germanium, Silicium, Granate) bekannte Erscheinung. Sie entstehen dadurch, daß das Wachstum an bestimmten Stellen auf „atomar glatten" Flächen stattfindet, während die Wachstumsfront im übrigen „atomar rauh" ist (vgl. WILKE). An einer (mikroskopisch) atomar rauhen Wachstumsfront ist die Unterkühlung vernachlässigbar klein; die Wachstumsfront folgt praktisch der Schmelzpunktsisothermen und hat dementsprechend (makroskopisch) einen gerundeten Verlauf. Dort hingegen, wo die Lage der Wachstumsfront mit einer morphologisch „wichtigen" (d. h. einer zur Gleichgewichtsform gehörenden) Kristallfläche übereinstimmt, kann es zur Bildung eines atomar glatten Bereiches, einer auch makroskopisch ebenen Facette kommen. Wegen der grundsätzlich anderen, gehemmten Wachstumskinetik an einer atomar glatten Fläche bleibt die Facette etwas hinter der Schmelzpunktisothermen bzw. hinter der atomar rauhen Wachstumsfront zurück, und vor der Facette besteht eine gewisse endliche Unterkühlung ΔT.

Schneidet man von einem zylindrischen bzw. kegelförmigen Kristallkörper mit dem Radius R eine Facette der Breite d ab, so gilt die geometrische Beziehung (Fig. 2):

$$(d/2)^2 = h(2R - h) \quad \text{bzw. für} \quad h \ll R \text{ auch: } d \approx 2\sqrt{2hR} \, .$$

Setzt man $\Delta T = hG_R$ mit G_R als radialem Temperaturgradient, so kann man auch $d \approx 2\sqrt{2R\,\Delta T/G_R}$ schreiben: Die (hinsichtlich der Züchtungspraxis unerwünschte) Facette wird um so breiter, je geringer der radiale Temperaturgradient G_R ist. Die Facetten erreichen eine Breite $d = 8 \ldots 10$ mm. Wie durch Messungen belegt, besteht in der Anfangsphase des Züchtungsvorganges beim Abschluß der Verbreiterung in dem

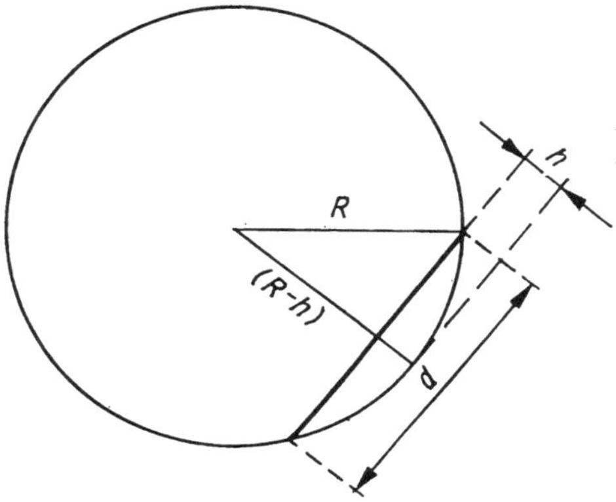

Fig. 2. Facette an einem zylindrischen bzw. kegelförmigen Kristallkörper

bis dahin gewachsenen Kristallkörper, der eine annähernd kegelförmige Gestalt gemäß Abbildung 4 hat, zwischen dem Ansatz des Keimkristalls und der Wachstumsfront eine Temperaturdifferenz von 25 ... 50 K. Insoweit man einen annähernd linearen Temperaturverlauf annehmen kann, ergibt sich bei einem Kristallradius $R = 2{,}5$ cm ein radialer Temperaturgradient $G_R = 10 \ldots 20$ K/cm. Das führt nach obigem Ausdruck auf eine Unterkühlung $\Delta T = 0{,}5 \ldots 1$ K vor der Facette. (Im weiteren Verlauf des Züchtungsvorganges werden mit dem Größerwerden des Kristallkörpers die Temperaturgradienten geringer und die Facetten größer.)

Tabelle 1
Facetten am Lithiumniobat

Flächen	1. $(01\bar{1}2)$	3. $(10\bar{1}\bar{2})$ 5. $(\bar{1}10\bar{2})$	2. $(1\bar{1}02)$ 6. $(\bar{1}012)$	4. $(0\bar{1}1\bar{2})$	(0001) $(000\bar{1})$
ϱ_y	32,75°	65,12°	114,88°	147,25°	90°
ϑ	147,25°	114,88°	65,12°	32,75°	90°

Nummer vor dem Flächensymbol gemäß Figur 1; ϱ_y Winkel zwischen Flächennormale und [12.0] (y-Achse); ϑ Winkel zwischen Fläche und Horizontale (Schmelzoberfläche) gemäß Figur 5.

Als morphologisch „wichtige", glatt wachsende und somit Facetten bildende Flächen treten am Lithiumniobat die der Form $\{01\bar{1}2\}$ (negatives Rhomboeder) und untergeordnet $\{0001\}$ (Basis) in Erscheinung. Die Flächennormalen schließen mit der y-Achse die in Tabelle 1 aufgeführten Winkel ϱ_y ein. An der Peripherie (Außenkante) der Wachstumsfront ändert sich die Richtung der Normalen auf die Kristalloberfläche bei einem stabförmigen Kristall um 90°, in der Phase der Verbreiterung des Kristalls sogar um ca. 150°, so daß hier auch Flächen der Oberseite des Kristalls (mit $\varrho_y < 90°$) Facetten bilden können. Infolge der anderen Wachstumskinetik, der Verformung des Schmelzmeniskus und der besonderen Benetzungsverhältnisse im Bereich der Facette verbleibt am Kristallmantel eine Spur in Form eines Kammes, welcher kristallographisch der durch (01$\bar{1}$0) und die Facettenfläche bestimmten Zone entspricht (vgl. Fig. 1). Die Kämme springen erhaben aus dem ansonsten gerundeten Umriß des Kristallkörpers hervor, was auf eine bevorzugte laterale Materialanlagerung an die Facettenfläche zurückzuführen ist (deren Mechanismus hier nicht diskutiert wird).

Insbesondere die Kämme an der Oberseite des Kristalls zeigen einen terrassenartigen Aufbau (Fig. 3). Das ist durch diskontinuierliche Bewegungen des Schmelzmeniskus während des Ziehvorganges zu erklären und weist auf Instabilitäten der Benetzungs-

Fig. 3. Kamm der Facette (10$\bar{1}$2) mit Polykristallbildung, Maßstab 1,5:1

verhältnisse hin. Während des Ziehens kann man direkt beobachten, daß der Schmelzmeniskus an der Facette gelegentlich um mehrere mm zurückspringt und etwas später wieder emporspringt. Dabei kann es zu der unerwünschten Bildung von Polykristallen kommen.

Meniskus an einem facettierten Kristall

An einem (nichtfacettierten) Kristallkörper mit kreisförmigem Querschnitt hat der Meniskus nach TSIVINSKI eine Höhe h_M (Fig. 4):

$$h_M = [a^2(1 - \sin\alpha) + a^4 \cos^2\alpha/16R^2]^{1/2} - a^2 \cos\alpha/4R$$

mit dem Kristallradius R und der Kapillaritätskonstanten $a = (2\gamma_{LV}/\varrho_L g)^{1/2}$ mit der Oberflächenspannung γ_{LV} und der Dichte ϱ_L der Schmelze; g ist die Erdbeschleunigung. SATUNKIN und ROSSOLENKO bestimmten experimentell aus der Beobachtung an Schmelzmenisken von $LiNbO_3$ folgende Werte: $\varrho_L = 3{,}80$ g cm^{-3}; $\gamma_{LV} =$ $= 300$ dyn cm^{-1}; $a = 0{,}40$ cm (bei 1270 °C). Der Wachstumswinkel $\varepsilon = \alpha - \beta$ wurde gleichfalls von SATUNKIN und ROSSOLENKO zu $\varepsilon = 0°$ beobachtet, d. h. es gilt $\alpha = \beta$. Bei einem zylindrisch wachsenden Kristall ist in diesem Fall $\alpha = \beta = 0°$. Diese Beobachtungen beziehen sich auf eine atomar rauhe Wachstumsfront, im folgenden durch einen oberen Index „r" gekennzeichnet. Für einen Kristall mit dem Radius $R = 2{,}5$ cm errechnet man eine Meniskushöhe $h_M^r = 0{,}38$ cm. In der Phase der Verbreiterung sind $\alpha = \beta > 0°$ und die Meniskushöhe geringer. So erhält man z. B. für $\alpha = \beta = 60°$ eine Meniskushöhe $h_M^r = 0{,}14$ cm.

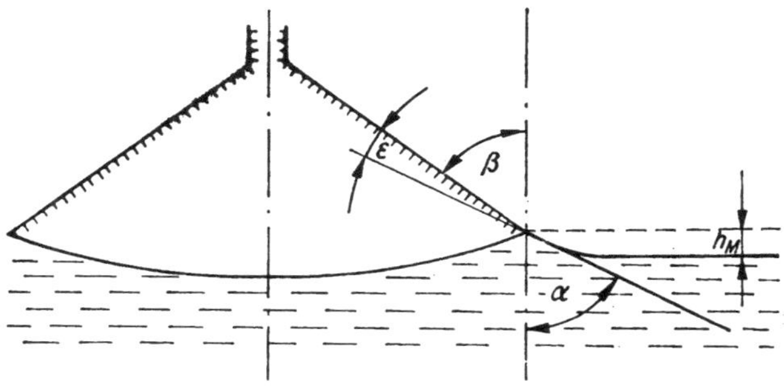

Fig. 4. Meniskus an einem kegelförmigen Kristallkörper

Ein Wachstumswinkel $\varepsilon = 0°$ ist jedoch nur kennzeichnend für Kristalle, die von ihrer Schmelze vollständig benetzt werden. In Übereinstimmung damit wird für Oxide im allgemeinen eine vollständige Benetzbarkeit durch ihre Schmelze angenommen. Diese Annahme dürfte jedoch nur für atomar rauhe Oberflächen zutreffen, deren Oberflächenspannung (spezifische freie Grenzflächenenergie) γ_{SV}^r wegen der ähnlichen atomaren Oberflächenstruktur ähnlich bzw. wenig größer als die Oberflächenspannung γ_{LV} vorausgesetzt werden kann: $\gamma_{SV}^r \gtrapprox \gamma_{LV}$. Aus demselben Grund dürfte die Grenzflächenspannung (spezifische freie Grenzflächenenergie) γ_{SL}^r zwischen einer atomar rauhen Fläche und der Schmelze sehr klein sein: $\gamma_{SL}^r \gtrapprox 0$. Das führt gemäß der Youngschen Beziehung:

$$\cos \Theta^r = (\gamma_{SV}^r - \gamma_{SL}^r)/\gamma_{LV}$$

auf einen Wert von $\cos \Theta^r \approx 1$ bzw. zu einem Kontaktwinkel $\Theta^r \approx 0°$; d. h. zu einer vollständigen Benetzung.

Hingegen hat eine atomar glatte, singuläre Kristallfläche eine markant kleinere spezifische freie Oberflächenenergie $\gamma_{SV}^g < \gamma_{SV}^r$, und die spezifische freie Grenzflächenenergie (Grenzflächenspannung) zwischen einer glatten Fläche und der Schmelze ist deutlich größer: $\gamma_{SL}^g > \gamma_{SL}^r$, so daß hier ein endlicher Kontaktwinkel $\Theta^g > 0$, d. h. eine unvollständige Benetzung zu erwarten ist. Da jedoch für die singulären Kristallflächen von $LiNbO_3$ weder γ_{SV}^g noch γ_{SL}^g bekannt sind, kann man Θ^g nicht unmittelbar berechnen.

Für eine Abschätzung sei einmal $\gamma_{SV}^g = \gamma_{LV}/3 = 100$ dyn cm^{-1}, zum anderen $\gamma_{SV}^g = \gamma_{LV}/2 = 150$ dyn cm^{-1} angenommen. Die Abschätzung der Grenzflächenspannung γ_{SL}^g sei sowohl nach der Regel von Antonov: $\gamma_{SL}^g = \gamma_{LV} - \gamma_{SV}^g$, als auch nach der Regel von Fowkes: $\gamma_{SL}^g = \gamma_{LV} + \gamma_{SV}^g - 2\sqrt{\gamma_{LV}\gamma_{SV}^g}$ vorgenommen (Tab. 2).

Tabelle 2
Abschätzung des Kontaktwinkels Θ^g für $\gamma_{LV} = 300$ dyn cm^{-1}

γ_{SV}^g/dyn cm^{-1}	γ_{SL}^g/dyn cm^{-1}		Θ^g	
	Antonov	Fowkes	Antonov	Fowkes
100	200	54	109°	81°
150	150	26	90°	66°

Da es sich nur um sehr grobe Abschätzungen handelt, soll die Akkuratesse dieser Regeln hier nicht diskutiert werden; auch wird eine ebene, glatte Facettenfläche vorausgesetzt, und es wird nicht zwischen Gleichgewichts- und dynamischen Kontaktwinkeln unterschieden (vgl. hierzu Neumann). Diese Abschätzung liefert gemäß

Tabelle 2 bemerkenswert große Kontaktwinkel im Bereich von $\Theta^g = 65 \ldots 110°$. Mit diesen Kontaktwinkeln werde der kapillare Anstieg h_F^g der Schmelze an einer um einen Winkel ϑ gegen die Horizontale geneigten ebenen Fläche (Fig. 5) gemäß $h_F^g = a[1 + \cos(\Theta + \vartheta)]^{1/2}$ (NEUMANN) berechnet (Tab. 3). Bei den Facetten $(1\bar{1}02)$ und $(\bar{1}012)$ beträgt der Neigungswinkel $\vartheta = 65{,}12°$ und bei den (für die Polykristallbildung kritischen) Facetten $(10\bar{1}\bar{2})$ und $(\bar{1}10\bar{2})$ ist $\vartheta = 114{,}88°$. Für diese letzteren Facetten an der Oberseite ergibt sich anstelle eines kapillaren Anstiegs eine kapillare Depression, und der Unterschied in der Schmelzhöhe beträgt $\Delta h = h_F^g - h_M^r = -0{,}14 \ldots -0{,}36$ cm. Beim Ziehen hat die Schmelze zunächst rings um den Kristall die Höhe h_M^r. Wenn sich die Facette ausbildet, bleibt der Meniskus zunächst an ihrem oberen Rand „angeheftet" und wird beim Ziehen etwas mit angehoben. Mit zunehmender Größe der Facette wird diese Meniskuskonfiguration instabil, so daß die Schmelze vom oberen

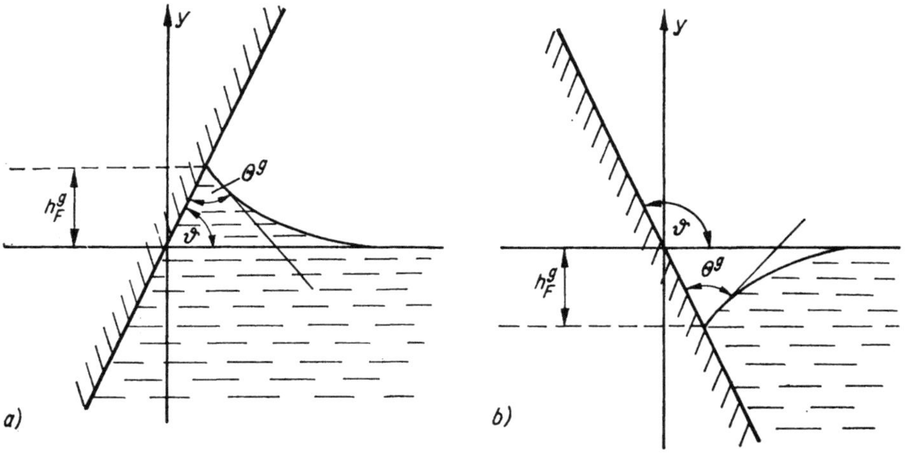

Fig. 5. Kapillaranstieg an geneigten Flächen. a) Facette an der Unterseite, b) Facette an der Oberseite

Tabelle 3
Berechneter Kapillaranstieg an Facettenflächen

Θ^g	$(10\bar{1}\bar{2})$; $(\bar{1}10\bar{2})$			$(1\bar{1}02)$; $(\bar{1}012)$			$(0\bar{1}1\bar{2})$		
	h_F^g/cm	Δh/cm	Δl/cm	h_F^g/cm	Δh/cm	Δl/cm	h_F^g/cm	Δh/cm	Δl/cm
65°	+0,00	−0,14	−0,15	0,24	+0,10	0,11	0,37	0,23	0,43
90°	−0,12	−0,26	−0,29	0,12	−0,02	−0,02	0,27	0,13	0,24
110°	−0,22	−0,36	−0,40	0,02	−0,12	−0,13	0,18	0,04	0,07

Rand der Facette abreißt und zurückspringt, um den Kontaktwinkel Θ^g herzustellen. Insoweit die Facette groß genug ist, springt dann die Schmelzoberfläche auf der Facette um eine Länge $\Delta l = \Delta h/\sin \vartheta = -0{,}15 \ldots -0{,}40$ cm gegenüber der Umgebung zurück. Eine solche schmale Depression der Schmelze an der Facette ist sowohl hydrostatisch als auch bezüglich der Oberflächenspannung recht labil, und bei einer (z. B. durch die Rotation des Kristalls bedingten) Störung flutet die Schmelze wieder an der Facette empor. Da sich die Facette inzwischen abgekühlt hat, kann dabei der unerwünschte Effekt eintreten, daß die auf die Facette zurückflutende Schmelze spontan kristallisiert und Polykristalle bildet.

Für die Facetten (1$\bar{1}$02) und ($\bar{1}$012) an der Unterseite ist die Situation weniger kritisch: Die Differenz der Schmelzhöhen errechnet sich zu $\Delta h = h_F^g - h_M^r =$ $= +0{,}10 \ldots -0{,}12$ cm; d. h. man erhält je nach den für die Abschätzung angenommenen Werten von γ_{SV}^g und γ_{SL}^g entweder einen geringen Kapillaranstieg oder eine geringe Kapillardepression (wobei auch $\Delta h = 0$ im Rahmen der Abschätzung liegt); auch die Verformung der Schmelzoberfläche bleibt mit $\Delta l = +0{,}11 \ldots -0{,}13$ cm entsprechend klein. Zudem kühlt sich die nach unten zur Schmelze gerichtete Facette, wenn sie freiliegt, nicht so stark ab, so daß die Gefahr der Polykristallbildung nicht besteht.

Bei der (unteren) Facette (0$\bar{1}$1$\bar{2}$) erhält man stets einen Kapillaranstieg. Bei der (oberen) Facette (01$\bar{1}$2) schließlich unterbleibt das Zurückspringen der Schmelze, da sie sich nicht unter dem Schmelzmeniskus fortsetzt.

Einwirkung eines elektrischen Feldes

Die Züchtung von $LiNbO_3$ unter Einwirkung eines elektrischen Feldes ist seit längerer Zeit bekannt und wird für die Präparation des Eindomänenzustandes sowie zur Beeinflussung von Dotierungen benutzt (RÄUBER). Ein elektrisches Feld zwischen Kristall und Schmelze beeinflußt außerdem die Morphologie der wachsenden Kristalle. Besonders interessant ist, daß auch die Bildung von Polykristallen in der Phase der Verbreiterung unterdrückt wird (RUDOLPH et al.). Das elektrische Feld wird in der Weise angelegt, daß ein Strom vom Tiegel (positiv) über die Schmelze in den Kristall (negativ) geleitet wird. Dabei wird der Strom allmählich gesteigert, so daß die Stromdichte an der Grenzfläche Schmelze/Kristall 1...10 mA/cm² beträgt. Eine umgekehrte

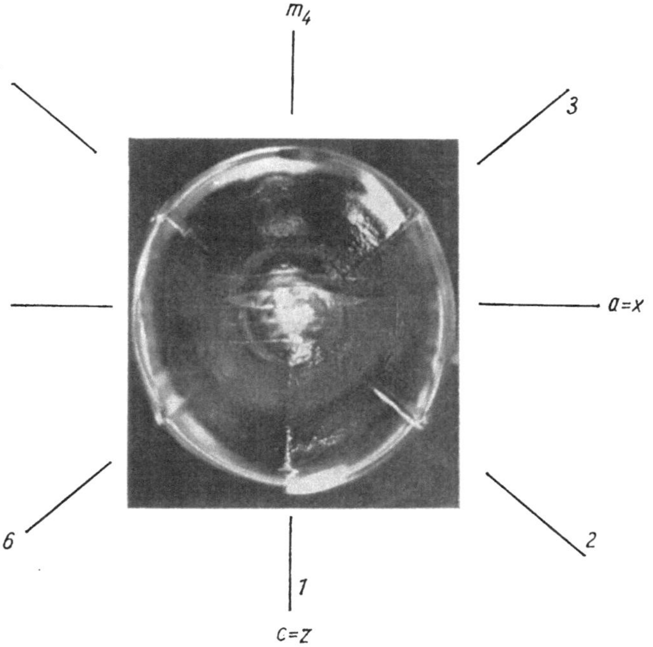

Fig. 6. Kopfbild eines $LiNbO_3$-Kristalls, gezogen in Richtung [12.0] (y-Achse) unter Einwirkung eines elektrischen Feldes. Strom: 25 mA, Tiegel positiv (Bedeutung der Ziffern wie in Figur 1), Maßstab 1:1,5

Polung (Kristall positiv, Tiegel negativ) führt infolge elektrolytischer Erscheinungen zur Blasenbildung im Kristall und verbietet sich deshalb.

Der Potentialsprung an der Grenzfläche läßt sich nur schwer abschätzen. Der Spannungsabfall über die gesamte Kristallänge (100 mm) liegt in der Größenordnung von 10 V.

Durch die Einwirkung des elektrischen Feldes wird die Oberflächenstruktur der gezüchteten Kristalle deutlich verändert (Fig. 6). Im zylindrischen Teil entstehen am Kristallmantel (bei Züchtung in y-Richtung) glatte {0001}-Facetten. Außerdem verändert sich der schwach elliptische Kristallquerschnitt in der Weise, daß die Richtung [00.1] (z-Achse) zum größten Durchmesser wird.

Die Unterbindung der Polykristallbildung an den Facetten läßt sich durch den Effekt der Elektrokapillarität erklären. Zufolge dieses Effektes wird durch den Potentialsprung an der Grenzfläche die Grenzflächenspannung γ_{SL}^g vermindert (vgl. KORTÜM). Eine Verminderung von γ_{SL}^g führt nach der Youngschen Beziehung zu einer Verkleinerung des Kontaktwinkels Θ^g, d. h. die Benetzung der Facetten wird verbessert. Dadurch wird die Kapillardepression an den oberen Facetten vermieden, und die oben geschilderten Vorgänge im Zusammenhang mit der Instabilität der Meniskuskonfiguration unterbleiben. Das wird durch die Beobachtung bestätigt.

Durch den Effekt der Elektrokapillarität wird auch an der atomar rauhen Grenzfläche die Benetzung verbessert bzw. der Spreitungsdruck verstärkt, worauf die veränderte Oberflächenstruktur zurückzuführen ist. Schließlich ist auch ein Effekt des elektrischen Feldes auf die Wachstumskinetik sowohl an atomar rauhen als auch an atomar glatten Grenzflächen nicht auszuschließen. So bleiben die Facetten unter Einwirkung des Feldes schmaler, was eine geringere kinetische Unterkühlung ΔT bedeutet.

Literatur

ANTONOV: siehe NEUMANN
FOWKES, F. M.: siehe NEUMANN
KORTÜM, G.: Lehrbuch der Elektrochemie, Weinheim/Bergstr. 1957
NEUMANN, A. W.: Adv. Colloid Interface Sci. **4** (1974) 105
NIIZEKI, N., YAMADA, T., TOYODA, H.: J. Appl. Physics **6** (1967) 318
RÄUBER, A.: Chemistry and physics of lithium niobate; in: Current Topics Materials Science, Vol. 1, S. 481–601, E. KALDIS (ed.); Amsterdam, New York, Oxford 1978
RUDOLPH, P., REICHE, P., SCHULTZE, D., HERMONEIT, B.: DDR-Patent Nr. 235890 A1, 21. 05. 86
SATUNKIN, G. A., ROSSOLENKO, S. N.: Cryst. Res. Technol. **21**, (1986) 1125
TSIVINSKI, S. V.: Inženerno Fiz. J. **5**, (1962) 59
WILKE, K.-Th.: Kristallzüchtung, Berlin 1973

(Received September 21, 1987)

Author's addresses:

P. REICHE, Doz. Dr. habil. J. BOHM, B. HERMONEIT, Dr. D. SCHULTZE
Zentralinstitut für Optik und Spektroskopie der Akademie der Wissenschaften der DDR
Rudower Chaussee 5
Berlin-Adlershof
DDR-1199

Prof. Dr. sc. P. RUDOLPH
Sektion Physik der Humboldt-Universität Berlin, Wissenschaftsbereich Kristallographie
Invalidenstraße 110
Berlin
DDR-1040

Cryst. Res. Technol. | 23 | 1988 | 4 | 475–480

S. YANEVA, L. STOJANOVA, N. STOICHEV, Z. KAMENOVA

Institute for Metal Science and Technology of the Bulgarien Academy of Sciences
Sofia, Bulgaria

On the Mechanism of Al–Si Alloy Structure Modification by Metal Antimony

In the process of Al–Si alloys solidification in presence of Sb the intermetallic compound AlSb is taking obviously a significant part. An adsorptional incorporation mechanism of AlSb action is proposed. It depends strongly on the antimony concentration in the melt.

The modifying effect is influenced also by a possibility of nucleation and growth of the AlSb phase itself.

Bei dem Erstarrungsprozeß der Al–Si Legierungen in Gegenwart von Sb spielt offensichtlich die Intermetallische Phase AlSb eine Hauptrolle. Es wird ein Absorptionsmechanismus des Einbaus der AlSb-Phase für die Veredelungswirkung von Sb vorgeschlagen. Die Veredelungswirkung von Sb hängt stark von der Sb-Konzentration in der Schmelze ab.

Der Effekt der Veredelung läßt sich auch durch die Möglichkeit einer Keimbildung und des Wachstums der neuen AlSb-Phase vor der Kristallisationsfront beeinflussen.

Introduction

Nowadays Al–Si alloys are widely used in the casting industry due to their good foundry and mechanical properties. They strongly depend on the alloy structure and especially on that of eutectic silicon. Such elements as sodium, strontium and antimony can modify the silicon phase structure. Antimony has received much attention because of its permanent modifying effect. It is assumed by some authors that the modification influence of sodium is connected with the observed during the solidification supercooling which means that the freezing point of the eutectic decreases by about 10 K. The occurrence of such big supercoolings favours a reduction in silicon particles' dimensions because of the increasing number of silicon phase nuclei. This model can be useful for explanation of modification by sodium and partly by strontium. It is well known that the supercooling in presence of antimony is very small and the above mentioned model is not applicable to its influence explanation.

Many opinions have been offered about the antimony concentration, necessary for significant improvement of casting's mechanical properties (BORGEAUD et al.; BUDUROV et al.; GÜNTHER, JÜRGENS; JACOB). There are few data about the interaction of antimony and growing silicon phase (LÖHBERG, PATTANAYAK; YANEVA et al.) that gives no satisfying explanation of antimony distribution in the solidifying alloy and new additional experiments are necessary.

The subject of this study is to investigate the antimony distribution in the silicon particles and Al matrix of some Al–Si alloys and to give a new opinion on the antimony influence during their crystallization.

Experimental

The investigated alloys have been prepared by adding metal antimony to Al–Si melts. The alloys with Sb concentrations in the range of 0.06–0.6 wt% have been remelted and crystallized unidirectionally at rates of 2 to 12 cm/h. The antimony concentration has been determined on the corresponding cross sections by means of X-ray fluorescent analyser VRA-2. The microstructure of the investigated cross sections has been studied by metallographic microscope Reichert MeF-2. The microscale distribution of antimony among the phases has been investigated on the same specimens' cross sections by means of X-ray microanalyser JXM-733.

Results and discussion

The antimony appearance and distribution in silicon and aluminium phases as well as in phase boundaries has been established. Figure 1 envisions the incorporation of Al and Sb in silicon particles. The Sb content of the Si-phase in Al–7% Si alloy increases with increasing of the melt concentration (Fig. 2). The distribution coefficient $K_{eff} = C_{Si}/C_{melt}$ is significantly larger than the equilibrium distribution coefficient K_0

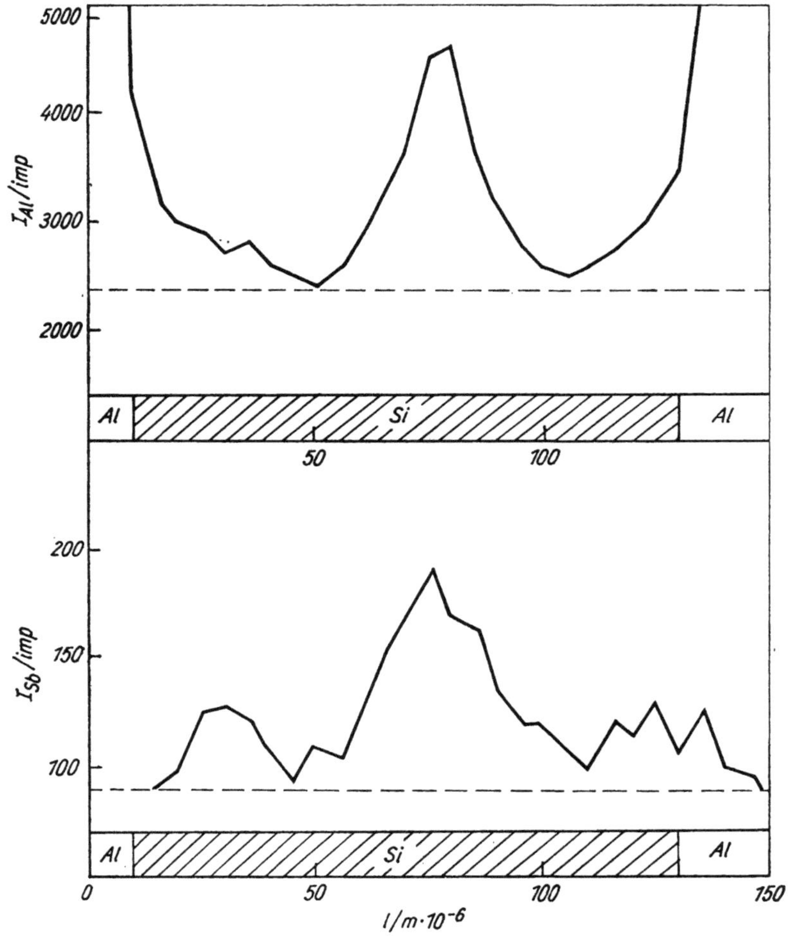

Fig. 1. Aluminium and antimony incorporation in a silicon particle (X-ray microprobe analysis of a cross section)

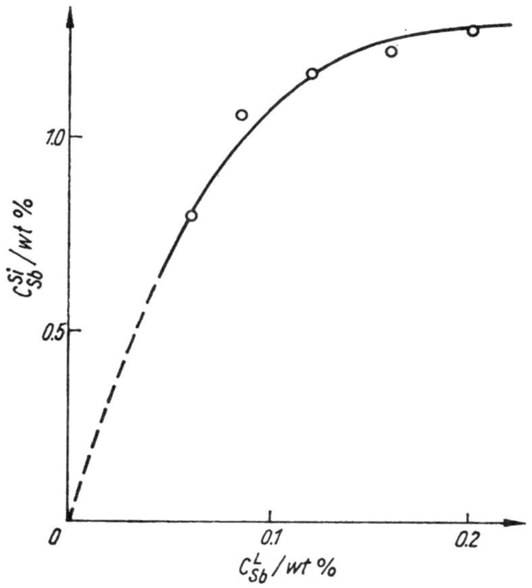

Fig. 2. Dependence of Sb inclusion in Si particles on the antimony concentration in the melt

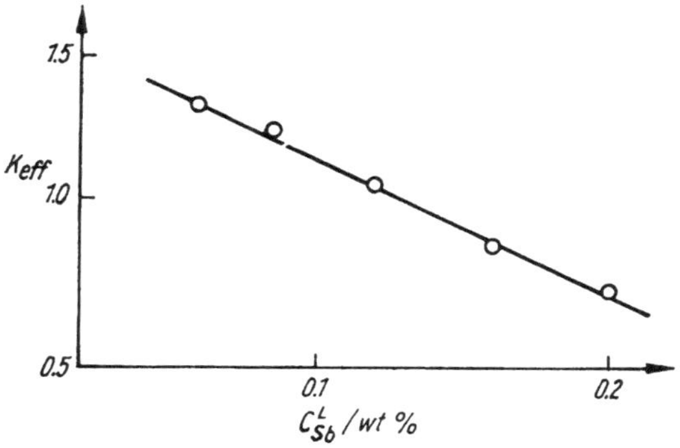

Fig. 3. Dependence of the effective distribution coefficient in hypoeutectic Al–Si alloys on the antimony concentration in the melt

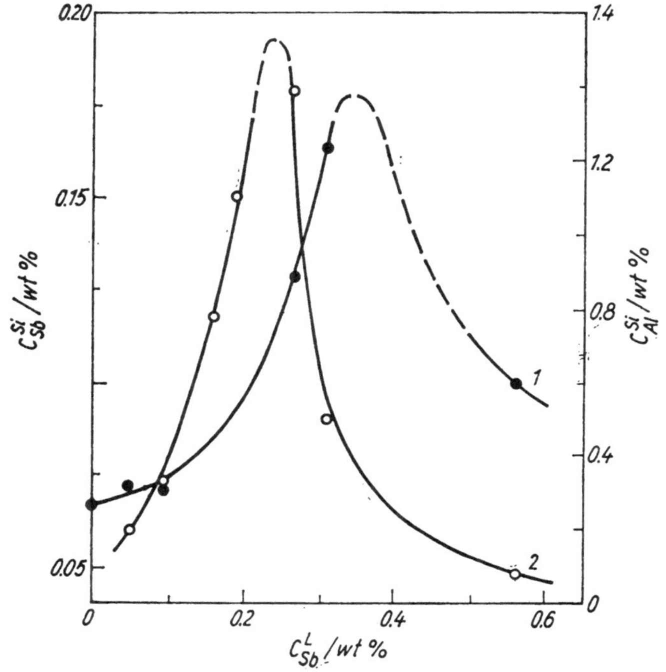

Fig. 4. Dependence of Al (1) and Sb (2) inclusion concentrations in the silicon phase on the antimony concentration in the melt

(according to GLASOV and ZEMSKOV at 1020 °C $K_0 = 0.0167$) and decreases with the increasing of the Sb concentration in the melt (Fig. 3). Precise quantitative investigation of the Al and Sb incorporation in the silicon particles has only been possible on the Si-phase larger than eutectic Si plates. So hypereutectic alloy with 17 wt% Si has been used. The distribution of Al and Sb has been studied in the primary Si crystals of that hypereutectic alloy. The concentration of both elements (Fig. 4) is highly sensitive of the Sb concentration in the melt. The incorporation of Al and Sb shows maximum at about 0.2—0.4 wt% Sb in the melt. Figure 5 shows the variation of the distribution coefficient K_{eff} with antimony concentration in the melt. The decreasing of the K_{eff} in the hypereutectic alloys has the same tendency as the hypoeutectic alloy.

According to MONDOLFO, in the aluminium corner of the ternary system Al–Si–Sb phase diagram there are three equilibrium phases: alpha-aluminium solid solution, silicon, and AlSb. GLASOV and ZEMSKOV show the increased dissolution of Al and Sb in Si when both components are in equimolecular quantities, due to chemical interaction of Al and Sb, forming AlSb, which is nearer to Si crystals by the type and dimensions of crystal lattice than either Al or Sb. The content of Al and Sb is higher than the equilibrium concentrations taken from the phase diagrams of the binary Al–Si and Si–Sb systems (according to GLASOV and ZEMSKOV eutectic concentrations are 0.12 at% Al in Si and 0.003 at% Sb in Si). Our data and those of other authors show clearly that the modifying effect is offered by AlSb.

In accordance with the phenomena in the salt systems crystallization (MERKULOVA et al.; JOFFE) it is assumed that the adsorption causes considerable changes in the kinetics of the Si-phase growth. The surface layer of growing silicon crystals is enriched by AlSb due to its adsorption, and that causes a decreasing of crystallization rate with increasing impurity concentration. The layer-by-layer mechanism of growth requires incorporation of impurity to depend on the ratio of the layer spreading rate to the desorption velocity. The layer of relatively high antimony concentration, which builds up on the growing surface, could be incorporated into silicon crystals by some circumstances. KRÖGER summarizes other authors evidence that when the adsorption takes part in the crystallization kinetics, there are some growth velocities when the distribution of impurities between solid and liquid phases corresponds to the adsorption equilibrium. If $K_{ads} > K_0$, the concentrations of impurities in the crystals can be higher than the equilibrium concentrations.

According to Hall's theory the distribution coefficient depends on the rate of crystal growth (V) and the desorption velocity (V_{des}):

$$K_{eff} = K_0 + (K_{ads} - K_0) \cdot \exp\left(-V_{des}/V\right) .$$

If $V \gg V_{des}$, the distribution coefficient becomes equal to K_{ads}. By fast growth V can reach and even becomes higher than V_{des} and that causes the incorporation of impurity layer which has not enough time to return back to the melt. According to JOFFE the distribution coefficient in this case depends on the impurity concentration in the melt and decreases with the increasing of the melt concentration. Figures 3 and 5 support the concept of an adsorption incorporation of AlSb in silicon crystals.

This decreasing of K_{eff} with increasing Sb concentration in the melt in connected with the effect of impurity adsorption on the crystal growth by assuming that a sufficient number of active sites for crystal growth exist on the crystal surface due to various reasons. These centres may be blocked by adsorption of impurity particles (AlSb molecules in this case) which causes a decreasing in the growth rate. In the extreme case all active sites will be blocked and AlSb will be rejected by the growing front. When the concentration of the AlSb in the melt reaches a critical value, there will be a real possibility for AlSb-phase to nucleate and grow. When that occurs, the modifying agent concentration will be effectively reduced. The microstructures of the alloys show coarser acicular silicon and the mechanical properties are severely affected.

It depends on the growth rate whether the AlSb particles will be incorporated or rejected by the growing silicon crystals. The dependence of Sb concentration in silicon crystals on the growth rate is given on Figure 6. It is assumed that nonequilibrium adsorption incorporation of AlSb in silicon crystals is connected with the high sensitivity of modification process of Sb in the Al–Si hypoeutectic alloys to the solidification conditions. In this way could be explained the influence of an overheating of the melt or an alteration of solidification rate of the castings to the modification process.

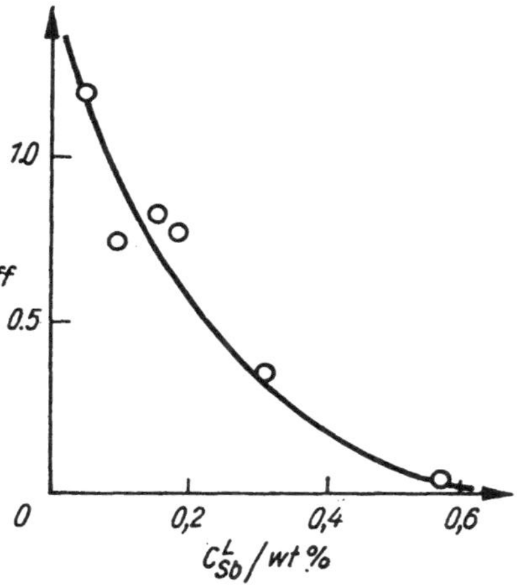

Fig. 5. Dependence of K_{eff} in hypereutectic Al–Si alloys on the Sb concentration in the melt

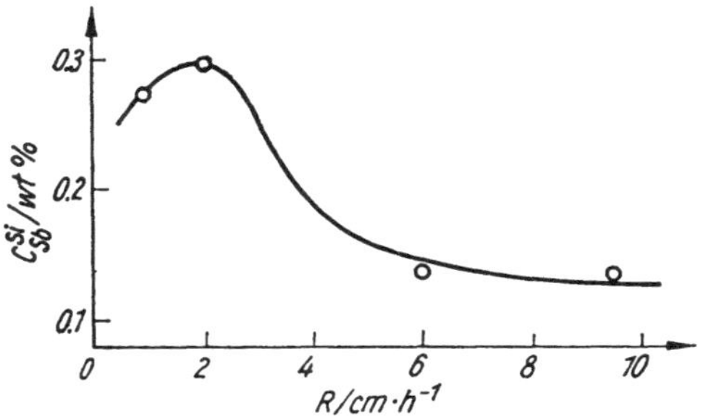

Fig. 6. Dependence of the antomony included in the silicon phase on the solidification rate

Conclusions

1. The modification of Al–Si alloys cast structure by Sb is due to the AlSb intermetallic compound that acts at the solidification front.
2. AlSb incorporates in the growing silicon phase by an adsorptional mechanism, depending strongly on the antimony concentration in the melt.
3. When the solidification conditions allow a new AlSb phase to nucleate and grow, the modifying effect is reducing significantly.

References

Borgeaud, P., Dabel, F., Drouzy, M., Mascrè, C.: Fonderie **269** (1967) 459
Budurov, S., Raikov, Chr., Yaneva, S., Stoichev, N., Russev, G.: Materials Science and Technology **11** (1982) 14
Glazov, N. M., Zemskov, V. C.: Fizikokhimitcheskie osnovi legirovania poluprovodnikov, Moscow 1967
Günther, B., Jürgens, H.: Giesserei **67** (1980) 8
Jacob, S.: Fonderie **363** (1977) 13
Joffe, E. M.: Radiokhimia **4** (1960) 381
Kröger, F. A.: The Chemistry of Imperfect Crystals, Amsterdam 1964

LÖHBERG, K., PATTANAYAK, D. R.: Aluminium **52** (1976) 368
MERKULOVA, M. C., POTAPOVA, S. A., SCHEBENKINA, G. S., RASTUCHIN, B. H.: Zhurnal Fizitcheskoi Khimii **5** (1957) 1075
MONDOLFO, L. F.: Aluminium Alloys: Structure and Properties, London–Boston 1976
YANEVA, S. B., STOICHEV, N. V., PANAJOTOVA, A. D.: Cryst. Res. Technol. **16** (1981) 737

(Received August 6, 1987
accepted October 2, 1987)

Authors' address:

Dr. S. YANEVA, Dr. L. STOJANOVA, Dr. N. STOICHEV, Dr. Z. KAMENOVA
Institute for Metal Science and Technology
Bulgarien Academy of Sciences
Chapaev Street 53
1574 Sofia
Bulgaria

Cryst. Res. Technol. | 23 | 1988 | 4 | 481–487

P. GILLE

Humboldt-Universität zu Berlin, Sektion Physik

On Segregation during Crystal Growth from a Solution Zone

The distribution in a mixed crystal $A_{1-x}B_xC$ grown from an either (A, B)-rich or C-rich solution zone (e.g. travelling heater method or related techniques) is to be described by PFANN's zone melting function which depends on the segregation coefficient and a characteristic length. Using pseudobinary mole fractions x the characteristic length of the distribution function is not the actual length of the liquid solution zone but an effective zone length is to be regarded which may be considerably smaller or much bigger for C-rich or (A, B)-rich solutions zones, respectively.

Die Verteilung in einem Mischkristall $A_{1-x}B_xC$, der aus einer (A, B)-reichen oder C-reichen Lösungszone gezüchtet wurde (z.B. travelling heater method oder ähnliche Verfahren), ist durch PFANNS Zonenschmelzfunktion zu beschreiben, die vom Verteilungskoeffizienten und der Zonenlänge abhängt. Wenn pseudobinäre Molenbrüche x zur Beschreibung verwendet werden, ist statt der tatsächlichen Länge der Lösungszone eine effektive Zonenlänge zu betrachten, die für den Verlauf der Verteilungsfunktion charakteristisch ist. Sie kann beträchtlich kleiner (C-reiche Lösungszone) oder wesentlich größer ((A, B)-reiche Lösungszone) als die wirkliche Länge der verwendeten flüssigen Zone sein.

Introduction

Besides different techniques to grow a crystal from a melt, crystal growth methods using some sort of solution have gained considerable attention with materials which cannot easily be crystallized from a stoichiometric melt. During the last twenty years, especially crystal growth from a solution zone has proved very successful in preparing a great variety of substances, because it combines the advantages of low temperature growth and steady state conditions. In fact, in steady state the dissolution of a source material just equals the crystallization from the solution zone. It is these conditions which make crystal growth from a solution zone particularly well accepted as a method of preparing mixed crystals.

The travelling heater method — sometimes better called "off-stoichiometric zone melting" — is the most employed solution zone technique in which the motion of the zone is produced by the motion of a heater. It has been successfully applied to the growth of single crystals of several semiconducting solid solutions such as (Hg, Cd)Te (TRIBOULET et al. 1985), (Cd, Zn)Te (TRIBOULET et al. 1983), (Hg, Zn)Te (TRIBOULET et al. 1986), (Pb, Sn)Te (GILLE et al.), (Ga, In)Sb, and (Ga, Al)As (YIP, WILCOX).

As it is described by many authors, the principle course of PFANN's segregation function is valid for a crystal grown from a solution zone, too:

$$c_s(z) = c_0 - (c_0 - c_L^0 k) \exp(-kz/l) . \tag{1}$$

In contrast to PFANN's original paper, the concentration of the starting molten zone c_L^0 is allowed to differ from that of the source material c_0. (k: segregation coefficient; z: axial position; l: length of the liquid zone.)

Dealing with segregation problems in mixed crystals it is usual to take the mole fraction x instead of the concentration c and to apply the segregation function in the same way. This is not just the right thing but a fairly good approximation because similar values of the mole volumes are one of the prerequisites which the constituents must meet to form a solid solution. Equ. (1) becomes to:

$$x_s(z) = x_0 - (x_0 - x_L^0 k) \exp(-kz/l) . \tag{2}$$

The axial distribution of the component under investigation is characterized by an initial transition region before the concentration level is asymptotically reached which is determined by the homogeneous concentration of the feed used.

In most cases it is allowed to assume the segregation coefficient to be constant over the concentration range involved. The exponent of the segregation function strongly depends on the characteristic length which is the length of the molten zone in zone melting experiments. That is why it is usual to express the grown axial distance in terms of the relative zone length z/l.

It is the aim of this paper to show that it cannot be applied in the same way to crystals grown from a solution zone as several authors do. When using pseudobinary mole fractions x to describe the distribution within a solid solution $A_{1-x}B_xC$ crystallized from an either (A, B)-rich or C-rich solution zone it is not correct to consider the actual length of the liquid zone to be the characteristic length of the segregation function, but the great amount of solvent is to be taken into account. For that reason a simple model is suggested which allows to calculate the characteristic length for the segregation in crystal growth from a solution zone which we call the effective zone length. It may be considerably smaller or much bigger than the actual zone length for C-rich or (A, B)-rich solutions zones, respectively.

Theoretic model

Let us consider a mixed crystal $A_{1-x}B_xC$ to grow from a solution zone $(A_{1-x_L}B_{x_L})_{1-y_L}C_{y_L}$ which may be either (A, B)-rich ($y_L < 0.5$) or C-rich ($y_L > 0.5$). It is convenient to use this pseudobinary description which better expresses that the segregation problem refers to the AC/BC ratio (i.e. mole fraction x) whereas the deviation from stoichio-

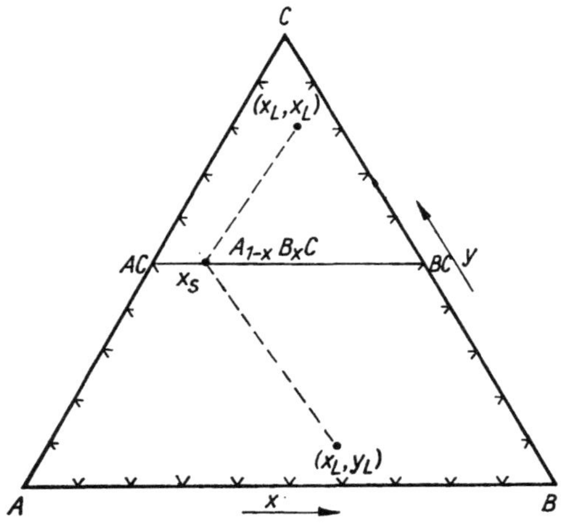

Fig. 1. Principle representation of a ternary phase diagram involving the pseudobinary solid solution $A_{1-x}B_x$ C. Dashed lines mark the equilibrium between a solid alloy with an (A, B)-rich or C-rich liquid phase (at the top read: (x_L, y_L))

metry in the liquid, which is the solubility y_L, is necessary for the solution growth mechanism (Fig. 1). PFANN's treatment of the segregation problem starts with the material's balance within the molten zone in terms of volume related units:

$$\frac{d(c_L \cdot l)}{dt} = v(c_0 - c_s) \tag{3}$$

(c: concentration of the component under consideration in the molten zone (c_L), in the source material (c_0), and in the grown crystal (c_s); l: length of the molten zone; v: translation rate of the two solid/liquid phase boundaries which are assumed to be flat; t: time).

Trying to write equ. (3) in terms of mole fraction x we see that it is not possible x_L simply to substitute for c_L, since c_L expresses the portion of the component under consideration within the ternary solution, i.e. also the solubility y_L, whereas x_L does not. If we like to compare the pseudobinary mole fractions in the solid and liquid solution (i.e. to use the pseudobinary segregation coefficient) we have to consider the dilution of the solution zone which is the amount of solvent used. In reality $(c_L \cdot l)$ depends on the portion of the solvent within the liquid zone. Instead of taking into account the change of c_L with varying solubility it seems more convenient to define a solubility-depending effective zone length l_{eff} which allows us to describe the material's balance in terms of mole fraction x for a solution zone, too:

$$\frac{d(x_L \cdot l_{eff})}{dt} = v(x_0 - x_s) . \tag{4}$$

l_{eff} covers all information about the actual zone length as well as the solubility y_L.

The effective zone length is that length which the liquid zone $(A_{1-x_L}B_{x_L})_{1-y_L}C_{y_L}$ would have, if it has had the same atomic fraction y as the solid (i.e. $A_{1-x_L}B_{x_L}C$). This

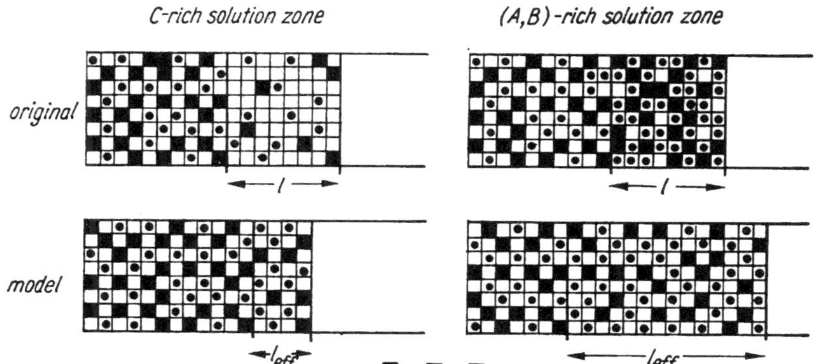

Fig. 2. Scheme of the solution zone arrangement being in equilibrium with a solid phase for a C-rich solution (on the left) and for an (A, B)-rich solution (on the right). The upper drawings represent the actual liquid zones, while the lower ones represent those liquid zones which are active with respect to segregation problems and which we call the effective zone lengths l_{eff}

means that the actual C-rich solution zone is to be imagined to decrease its amount of C up to the stoichiometric composition. For this reason the effective zone length of C-rich solution zones (more general: of solution zones with a solvent which remains neutral with respect to the components to segregate) is smaller than the actual one (Fig. 2, left part). On the other hand the (A, B)-rich solution zone (a solvent which consists of the components to segregate) must — in mind only — be filled with C to reach its effective length which is bigger than in reality (Fig. 2, rigth part).

Now, these effective zone lengths allow us to compare the mole fractions in the liquid and solid phases and to apply a pseudobinary segregation coefficient $k = x_s/x_L$ that can be taken from the phase diagram.

To put this idea into a suitable mathematical expression it is necessary to define the following values:

V_i molar volume of component i
x_L, x_s atomic fraction $N_B/(N_A + N_B)$ of the liquid and solid phase, respectively
y^L atomic fraction $N_C/(N_A + N_B + N_C)$ of the liquid phase.

For simplicity, we assume the molar volumes V_i to be independent of temperature and the total volume of the solution to be the sum of the individual volumes of the solvent and the solute. With a C-rich solution zone the actual zone consists of the effective (stoichiometric) zone and the excess of the pure solvent C:

$$(A_{1-x_L}B_{x_L})_{1-y_L} C_{y_L} \leftrightarrow (1 - y_L) A_{1-x_L}B_{x_L}C + (2y_L - 1) C; \qquad y_L \geqq 0.5 .$$

With the constant cross section which is the prerequisite for the whole derivation of the segregation function the ratio of the effective zone length to the actual one runs to:

$$\frac{l_{eff}}{l} = \frac{(1 - y_L) [(1 - x_L) V_{AC} + x_L V_{BC}]}{(2y_L - 1) V_C + (1 - y_L) [(1 - x_L) V_{AC} + x_L V_{BC}]}. \tag{5}$$

With the approximation $V_{AC} \approx V_{BC}$ which is fairly well met for the most solid solution systems, equ. (5) becomes to:

$$\frac{l_{eff}}{l} \approx \frac{1}{1 + \frac{2y_L - 1}{1 - y_L} \frac{V_C}{V_{AC}}}; \qquad y_L \geqq 0.5 . \tag{6}$$

In the case of crystals growing from an (A, B)-rich solution zone on the other hand, the actual liquid zone

$$(A_{1-x_L}B_{x_L})_{1-y_L} C_{y_L} \leftrightarrow y_L A_{1-x_L}B_{x_L}C + (1 - 2y_L) A_{1-x_L}B_{x_L}; \qquad y_L \leqq 0.5$$

has to be enriched with $(1 - 2y_L)$ C to take up its fictive volume according to the effective zone length. That is why, the ratio of the effective zone length to the actual one is greater than unity and runs to:

$$\frac{l_{eff}}{l} = \frac{(1 - y_L) [(1 - x_L) V_{AC} + x_L V_{BC}]}{y_L[(1 - x_L) V_{AC} + x_L V_{BC}] + (1 - 2y_L) [(1 - x_L) V_A + x_L V_B]}. \tag{7}$$

With the above mentioned approximation, which is in general not valid for the molar volumes V_A and V_B of the solvent, we get:

$$\frac{l_{eff}}{l} \approx \frac{1 - y_L}{y_L + \frac{1 - 2y_L}{V_{AC}} [(1 - x_L) V_A + x_L V_B]}; \qquad y_L \leqq 0.5 . \tag{8}$$

With the effective zone lengths defined in equ. (5) and (7) for crystal growth from a C-rich or (A, B)-rich solution zone, respectively, equ. (2) changes to:

$$x_s(z) = x_0 - (x_0 - x_L^0 k) \exp(-kz/l_{eff}) \tag{9}$$

which is valid for segregation during crystal growth from a solution zone, e.g. for the travelling heater method and related techniques.

Discussion

Figure 3 shows how the ratio of the effective zone length to the actual one depends on the atomic fraction y_L of the liquid zone. Values are taken from the Pb–Sn–Te system. This means, $y_L < 0.5$ belongs to a (Pb, Sn)-rich solution zone used for example in the (Pb, Se)Te growth by TSM (BANSARAGTSCHIN et al.) or by THM (JÄGER, SEIPP).

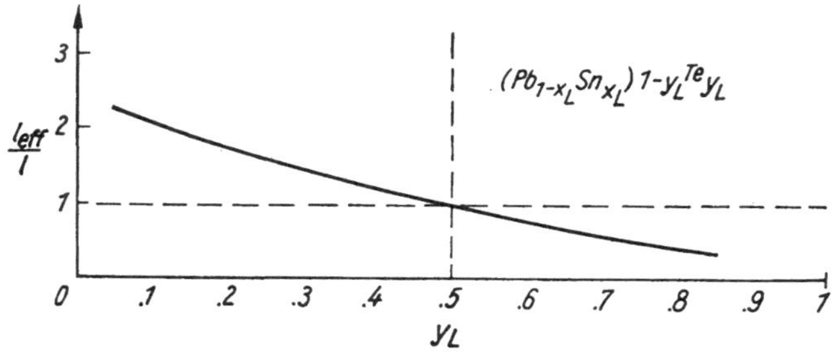

Fig. 3. Ratio of the effective (l_{eff}) to the actual (l) zone length versus the atomic fraction y_L of the liquid solution zone $(Pb_{1-x_L}Sn_{x_L})_{1-L}Te_{y_L}$, according to equ. (6) and (8) (read $(Pb_{1-x_L}Sn_{x_L})$ Te_{y_L})

With increasing solubility y_L (caused by an increasing temperature of the phase boundaries) the ratio l_{eff}/l diminishs and equals unity with $y_L = 0.5$, i.e. for a stoichiometric melt. This is the original case of stoichiometric zone melting. With a further increasing tellurium concentration (i.e. with a Te-rich solution zone) the effective zone length becomes smaller than the used liquid zone. This is to be observed in THM growth of (Pb, Sn)Te from Te-rich solutions zones (GILLE et al.) Apart from better growth conditions with Te-rich zones there is a considerable difference in the first part of the segregation functions of the two different solvents used.

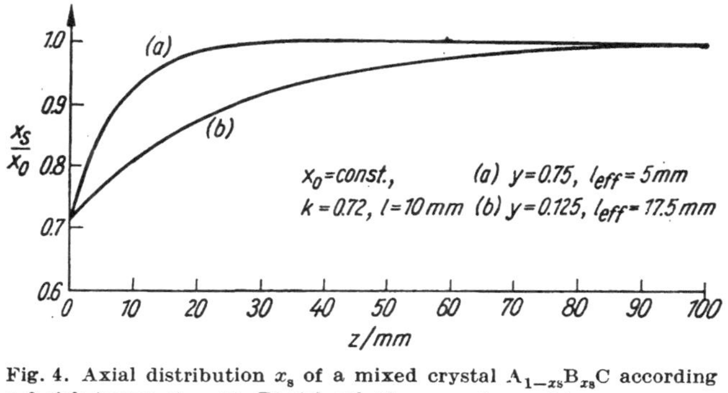

Fig. 4. Axial distribution x_s of a mixed crystal $A_{1-x_s}B_{x_s}C$ according to equ. (9) grown from a C-rich (curve a) or (A, B)-rich solution zone (curve b)

In Figure 4 the theoretic distribution x_s of a mixed crystal $A_{1-x_s}B_{x_s}C$ according to equ. (9) is pictured versus the axial position z for a C-rich solution zone (a) and an (A, B)-rich zone (b) with the actual zone length being equal in both cases. The conditions are chosen consistent with Figure 2 but differ from those of the Pb–Sn–Te system by taking the same segregation coefficients for the (A, B)-rich as well as the C-rich side to better demonstrate the great influence of the kind of solvent used. It is to be seen that with the aim of growing homogeneous crystals in solid solution systems it is much better to choose that component as the solvent which remains neutral with respect to segregation problems like in our example component C (i.e. Te for (Pb.

Sn)Te) does. It causes a relatively small effective zone length which is the characteristic length of the solution zone segregation function. Otherwise it will take the growing crystal a long distance to reach the level region of the distribution function, sometimes nearly as long as the total length of a THM-grown crystal.

Despite of all advantages of the solvent C of our example, it must be mentioned that this solvent acts as a relatively poor buffer when the growth conditions like the temperature or the homogeneity of the feed material are not as constant as they should be and as they are assumed in PFANN's derivation.

Experimental evidence

$Pb_{1-x}Sn_xTe$ crystals were grown by the travelling heater method using Te as the solvent. Details of the experimental procedure have been described elsewhere (GILLE et al.; GILLE). Use was made of a monocrystalline PbTe seed so that the $Pb_{1-x}Sn_xTe$ crystal grew with a constant cross section from the very beginning as it is supposed in the segregation function.

In Figure 5 the distribution of the mole fraction x in the first part of a $Pb_{1-x}Sn_xTe$ crystal is plotted versus the axial position. Mole fractions were measured by electron probe microanalysis (Sn–L_α radiation) along a longitudinal section of the crystal. It is to be seen that the growing crystal reaches the original composition of the source material after a distance of about 15 mm.

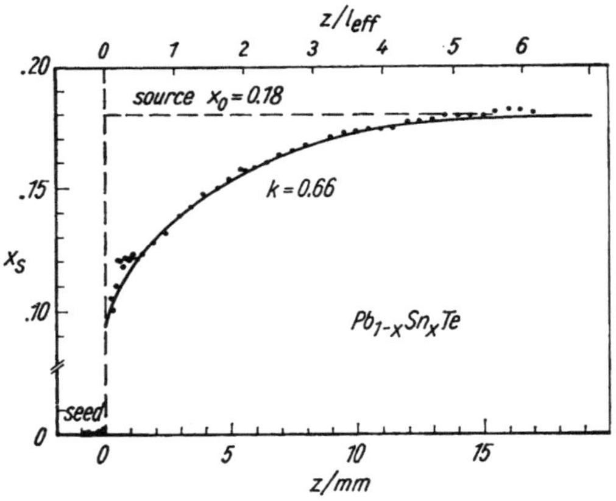

Fig. 5. Axial distribution x_s measured by electron probe microanalysis in a $Pb_{1-x}Sn_xTe$ crystal grown by the travelling heater method (THM) from an Te-rich solution zone

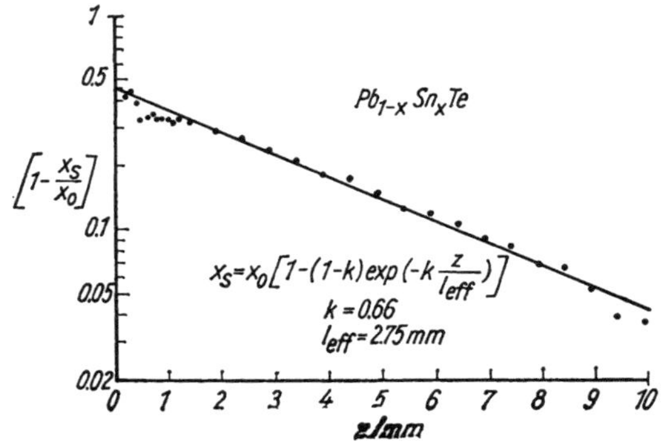

Fig. 6. Logarithmic representation of the axial distribution x_s in a $Pb_{1-x}Sn_xTe$ crystal grown by THM from a Te-rich solution zone (compare Fig. 5) revealing the segregation coefficient k and the effective zone length l_{eff}

When the distribution of the same crystal $(1 - x_s/x_0)$ is pictured in a logarithmic scale versus the axial position (Fig. 6) we get a linear course with the slope $(-k/l_{eff})$. From the intersection with $z = 0$, the segregation coefficient is determined to be $k = 0.66$, which is in very good agreement with the data known from WAGNER and WILLARDSON, and CALAWA et al. From this it follows an effective zone length of $l_{eff} = 2.75$ mm whereas the actual one was about 7 mm. This is the reason for the relatively fast convergency of the distribution function which allows the growth of homogeneous mixed crystals of (Pb, Sn)Te from a Te-rich solution zone by the travelling heater method.

The author is grateful to Dr. B. Christ for carrying out the electron probe microanalysis measurements.

References

BANSARAGTSCHIN, B., LINK, R., LEHMANN, G.: Kristall u. Technik **13** (1978) 269
CALAWA, A. R., HARMAN, T. C., FINN, M., YOUTZ, P.: Trans. Met. Soc. AIME **242** (1968) 374
GILLE, P., MÜHLBERG, M., PARTHIER, L., RUDOLPH, P.: Cryst. Res. Technol. **19** (1984) 881
GILLE, P.: Dissertation, Humboldt-Universität Berlin 1984
JÄGER, H., SEIPP, E.: Verh. DPG **1** (1978) 86
PFANN, W. G.: Trans. Met. Soc. AIME **194** (1952) 747
TRIBOULET, R., NEU, G., FOTOUHI, B.: J. Cryst. Growth **65** (1983) 262
TRIBOULET, R., NGUYEN DUY, T., DURAND, A.: J. Vac. Sci. Technol. A **3** (1985) 95
TRIBOULET, R., LASBLEY, A., TOULOUSE, B., GRANGER, R.: J. Cryst. Growth **79** (1986) 695
WAGNER, J. W., WILLARDSON, R. K.: Trans. Met. Soc. AIME **242** (1968) 366
YIP, V. F. S., WILCOX, W. R.: Mat. Res. Bull. **11** (1976) 895

(Received August 31, 1987
accepted September 12, 1987)

Author's address:

Dr. P. GILLE
Humboldt-Universität zu Berlin, Sektion Physik, Bereich Kristallographie
Invalidenstraße 110
Berlin
DDR-1040

Book Review

A. WEISS

Struktur und physikalische Eigenschaften metallorganischer Verbindungen

H. WENZL

Kristallzuchtforschung

Rheinisch-Westfälische Akademie der Wissenschaften, Vorträge N 353

Westdeutscher Verlag GmbH Opladen 1987, 64 Seiten, 21 Abbildungen, Preis DM 16,00, ISBN 3-531-08353-8

Das vorliegende Heft beinhaltet zwei Vorträge der 336. Sitzung der Rheinisch-Westfälischen Akademie der Wissenschaften vom 3. Dezember 1986. Beiden Niederschriften sind Literaturquellen sowie die Protokolle der Diskussionen beigefügt, was eine gute Ergänzung der Vortragsinhalte darstellt.

Im ersten Beitrag wird von A. WEISS ein komprimierter Überblick über den Stand der Struktur- und Eigenschaftsforschung auf dem modernen Gebiet metallorganischer Festkörper gegeben. Im Mittelpunkt stehen die intermolekulare und intramolekulare Geometrie und Dynamik der Verbindungen vom Typ $MIV(CH_3)_4$, wie z. B. $Sn(CH_3)_4$. Zur vollständigen Klärung des Bindungscharakters werden klassische thermodynamische Meßmethoden sowie Raman-, IR-Spektroskopie, magnetische Kernresonanz und Neutronenstreuung eingesetzt. Es deuten sich Ordnungs-Unordnungsphänomene und stufenförmige Übergänge in die flüssige Phase an. Es wird der interdisziplinäre Fundus dieses Forschungsgebietes für Strukturanalytiker, Physiker und Chemiker deutlich; offen bleibt dagegen die Frage der Anwendung.

Im zweiten Beitrag wird von H. WENZL eine anschauliche Übersicht über substanzübergreifende Probleme der Kristallzüchtung gegeben. Die Aussagen sind allgemeiner Art und werden von den instruktiven Ergebnissen der Jülicher Forschergruppe, so u. a. von vier exzellenten Fototafeln gestützt. Abgehandelt wird prägnant der gesamte Wachstumsprozeß und seine Grundfragen, wie Keimbildung, freie Kristallisation (Dendritenwachstum), erzwungene Kristallisation, kooperatives Wachstum (Eutektika und Peritektika) und Kristallfehler. Schwerpunkte bilden das Czochralskiverfahren, Konzentrationsunterkühlung und die Ursachen für die Entstehung periodischer Inhomogenitäten im Kristall. Dazu werden bildhafte computersimulierte Strömungsfelder in Czochralski-Züchtungstiegeln vorgestellt. Die Abhandlung endet mit einem Blick auf zukünftige Aufgaben der Kristallzüchtung und einer geradezu philosophischen Huldigung an die Kristallzüchter.

Dieses Heft ist den Laboratorien und Studieneinrichtungen für Materialforschungen zu empfehlen.

P. RUDOLPH

Cryst. Res. Technol. | 23 | 1988 | 4 | 489–497

O. P. FEDOROV, D. E. OVSIENKO

Institute of Metal Physics, Ukrainian Academy of Sciences, Kiev, USSR

Mechanism of Dendritic Side-Branches Coarsening under Isothermal Conditions

The process of dendritic branches transformation of succinonitrile and some other organic substances has been studied under the isothermal holding of two-component melt. The dendritic branches have been shown to be separated from the steam in the specimen with the thickness comparable with the diameter of the dendrite. Dendritic structure coarsening observed in the bulk specimen is not associated with the separation of the branches. In the region of the concentration gradient the dendritic growth has been shown to result in partial dissolution of as-crystallized material under the isothermal holding after the end of crystallization. This effect seems to result from the diffusionless growth in the two-phase region of the phase diagram.

Исследован процесс преобразования ветвей дендритов сукцинонитрила и некоторых других органических веществ при изотермической выдержке двухкомпонентного расплава. Показано, что отделение ветвей от ствола происходит только в препарате, толщина которого сопоставима с диаметром дендрита. В объемном препарате наблюдается укрупнение дендритной структуры, не связанное с отделением ветвей. Показано, что в результате роста дендритов в области концентрационного градиента часть закристаллизовавщегося объема растворяется при изотермической выдержке после окончания кристаллизации. Этот эффект связывается с бездиффузионным характером роста в двухфазной области диаграммы состояния.

Introduction

It is known that besides dendritic crystal growth the coarsening of the dendritic branches takes place. Based on the observations of the transparent models the driving force of the coarsening is believed to be a surface energy and this process may run along one of three mechanisms (FLEMINGS).

For the whole three models the calculation schemes are identical: on the basis of the equilibrium conditions between the crystal and the melt a diffusion flow of the dissolved substance, J, directed to a tip of dendritic branch with the radius r from a neighbouring section having smaller curvature $(1/\tilde{r})$ is written

$$J = (D/r)\,[C_s(r) - C_s(\tilde{r})]\,, \tag{1}$$

where D is the diffusivity, C_s is the concentration depending on a curvature of a dendritic section. As a result of the diffusion a melting temperature of more curved section is lowered and it is dissolved. So, for the model II the dissolution rate has been obtained (KAHLWEIT)

$$\frac{dl}{d\tau} = \frac{2D\sigma V^2 C_\infty}{kT} \cdot \frac{1}{r^2} \tag{2}$$

where l is the length of the branch, V is the molar volume, σ is the surface energy, C_∞ is the mean concentration of the impurity, r is considered to be constant.

Schemes I and II (FLEMINGS) suggest the isothermal coarsening of the large branches on account of dissolution of small ones. Scheme III is realized due to sidebranches separation from a dendritic steam followed by their dissolution.

The various mechanisms of the dendritic structure transformation should affect differently on a cast structure. Therefore, the study of the application features of above mechanisms is significant in principle.

It would be noted that various authors' data are inconsistent. It has been observed (KLIJA) the dendritic branches separation in model alloy under the isothermal conditions. It has been shown (JACKSON et al.) that the branches are separated from the dendritic steam with temperature increase only, i.e. due to melting of impurity enriched sections. The aim of our work is to study experimentally the dendritic branches transformation in transparent organic alloys under isothermal condition.

Experimental technique

A few systems studied before, e.g. cyclohexanol-water, ammonium chloride solution, tetrabromomethane-salol as well as succinonitrile-camphene and succinonitrile-salol have been chosen.

The specimen under study was placed in plane-parallel sealed glass vessel with thickness from 0.1 to 3 mm. Besides, the vessels with 2—3 cm thickness specimen were used to study a crystallization of ammonium chloride solution in a large volume.

The vessels with a plane specimen were held at constant temperature with the help of U-10-type thermostat with the precision of 0.01 K. The temperature was measured by the thermometers placed in an inlet and outlet of thermostated cell. A difference in indication of the thermometers did not exceed 0.01 K.

In a number of experiments the specimen under study was placed between a silver mirror and a cover glass. In this case the sample was thermostated on the both sides. The growth and the following transformation of the dendritic structure was registered by microphotography.

Results and discussion

Bulk specimen

For the samples where a vessel size is many times more than the diameter of the dendrite coarsening of the dendritic branches being not associated with their separation from a steam was observed. This fact is illustrated in Figure 1 where thin neck is artificially formed by partial melting of the crystal. In time, as one can see, solid is redistributed from the tip of the branch to its neck, i.e. dendritic shape becomes a sphere. This picture is typical for the branches grown sufficiently far from each other.

When spacings between the dendritic branches are very small, i.e. when l is much larger than the distance between branches, a, coarsening is determined to a high extent by coagulation of the branches. As seen in Figure 2 the short branches near the tip of the dendritic steam are smoothed out, but long ones grown far away from the tip coagulate. The mode of coagulation is determined by the relation l/a. When $l/a \lessapprox 1$ a hill-like structure is formed, with cavities smoothing in time. If $l/a \gg 1$ the neighbouring branches begin to coagulate in their tip range. Let us consider the modes of transformation of the branches at various l/a with the help of the scheme in Figure 3. At $l/a < 1$ (Fig. 3a) the dendritic branches may be considered as small protrusions on the steam surface. Smoothing out of such protrusions is well correctly described

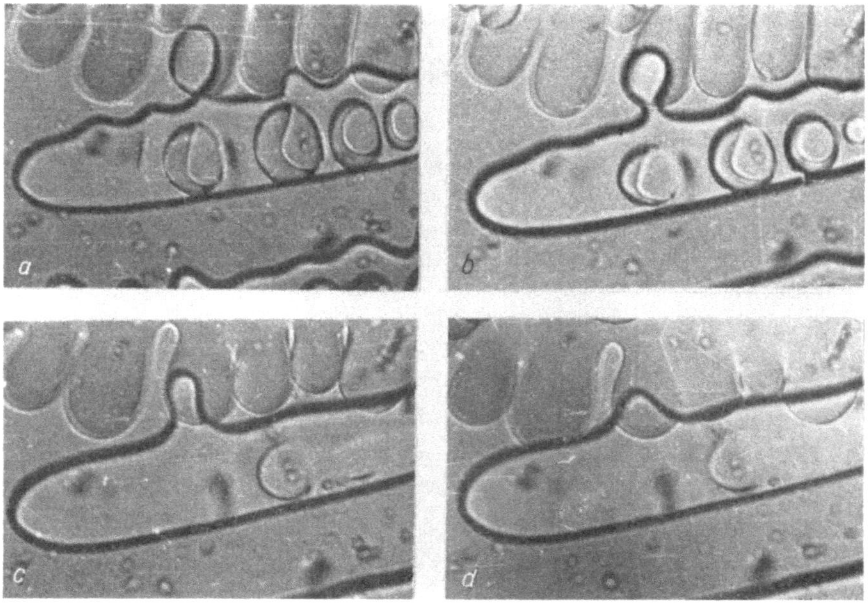

Fig. 1. Transformation of dendritic neck in succinonitrile-salol alloy: a) 0, b) 1, c) 3, d) 6 hours of isothermal holding

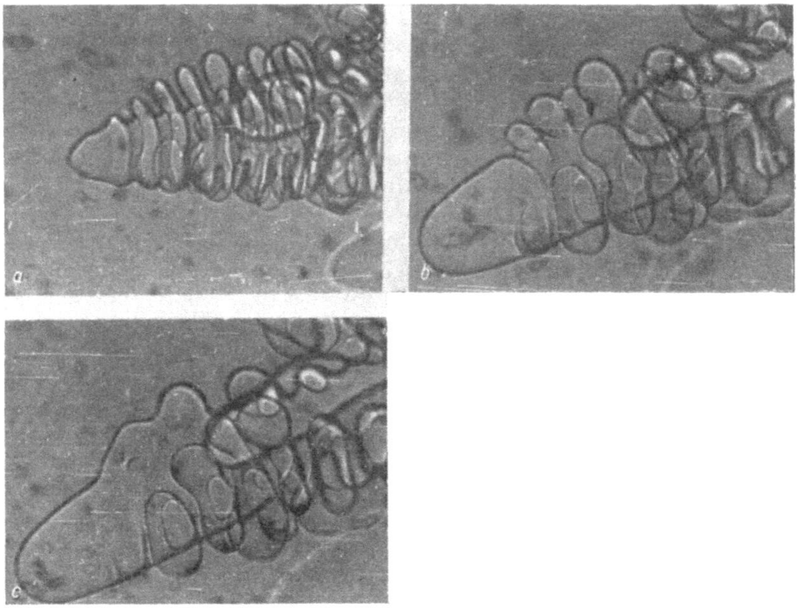

Fig. 2. Coarsening of dendrites in succinonitrile-salol alloy under isothermal holding at $T_s < T < T_l$: a) 1 min, b) 2 and c) 4 hours

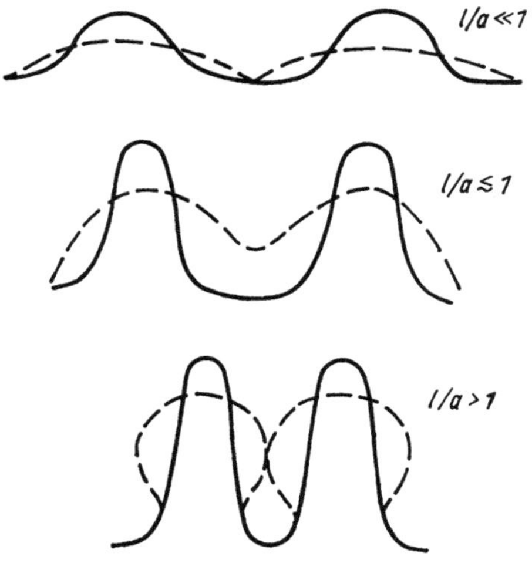

Fig. 3. Schemes of dendritic structure transformation depending on relation between l and a

by the Kahlweit model. The intermediate case $l/a \lesseqgtr 1$ is the most complicated one for an estimation. According to eq. (1) a flow of solid redistributed from one surface section to other is determined both by the impurity concentration difference and by a distance of mass transfer. In the case under consideration redistribution of solid from a tip to side surface of the branch becomes important too. Therefore the radius of the tip increases in time that results in decreasing of dissolution rate of the dendritic branch. Besides, the coagulation of the neighbouring branches becomes possible. It would be noted here that the branches are irregularly placed on the steam, as a rule. This irregularity results from fast move of "active" branches and from their suppression of "inactive" ones at a growth stage of the dendrite (HUANG, GLICKSMAN). During isothermal holding two neighbouring branches are coagulated, new branch forming with larger radius of the tip than initial ones. This new branch dissolutes slower than noncoagulated ones. The impression may be formed that scheme I (FLEMINGS) is realized. However, the experiments carried on systems of succinonitrile-salol and ammonium chloride solution reveal that both branches are dissolved under isothermal conditions, i.e. this process is similar to the KAHLWEIT scheme but it's more complex in comparison with this scheme due to altering radius of the branch tips. And, finally, in the last case $l/a \gg 1$ a redistribution of solid from the branch tip to its side surface becomes a decisive process. As a result the branch near its tip is thickened and the neighbouring branches are coagulated.

The above modes of the dendritic structure transformation have been studied in detail in succinonitrile-salol alloys (2.5%, 10% and 15% of salol). In other systems studied here: succinonitrile-camphene, succinonitrile-camphor, tetrabromo-methane-camphene, cyclohexanol-water, ammonium chloride-water the dendritic branches have not been separated during isothermal holding in the bulk specimen. The above features are typical also for the dendritic structure transformation of pure succinonitrile under isothermal conditions (VOORHEES, GLICKSMAN). We shall note that above systems with various parameters of the phase diagrams show qualitatively similar picture of the dendritic branches transformation. It gives reason to draw unambiguous conclusion about the dendritic structure transformation during isothermal holding. Namely, in these systems in absence of stir, convection, chemical inhomogeneity the dendritic structure is coarsened due to redistribution of solid from more curved surface sections to less curved ones. This process is not accompanied by the dendritic branches separation from the steam.

Thin specimens

For thin specimens with thickness (0.1 mm) compared with a diameter of growing dendrite a picture of the branch transformation is essentially differed from that for the bulk ones. For above systems the dendritic branches were separated from a steam during isothermal holding. A plane vessel with variable thickness (0.05 to 0.5 mm) was made to determine the vessel thickness influence on the process under study. The homogeneity of the components distribution was verified by slow melting of the sample after charge of the vessel with succinonitrile-salol alloy. The degree of homogeneity has been found to be not lower than in usual plane-parallel vessel. During isothermal holding between T_l (liquidus) and T_s (solidus) transformation with the separation has been observed in a thin part of the vessel but not in a thick one. The middle part of the vessel is shown in Figure 4. A mode of an attachment of the dendritic branches to the steam is changed from right: to left from remelted thin necks to unremelted ones and then to thickened ones. Thus, depending on a thickness of the vessel the various modes of the dendritic structure transformation are obtained with the branches separated from a dendritic steam or without separation.

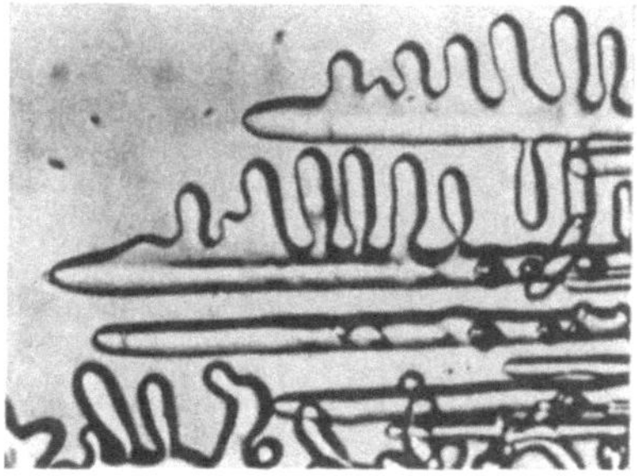

Fig. 4. The middle part of wedge-like vessel with succinonitrile-salol alloy

Let's consider the processes responsible for the dendritic structure transformation in thin specimens in more detail. As follows from the experiments carried on the various systems a part of solid is noticeably decreased during isothermal holding. It may be judged about this by an area occupied by "the plane dendrites" with respect to a total area of the specimen. The same conclusion follows from the analysis of (Klija). Abovementioned dissolution of solid is not resulted from temperature fluctuation: as special experiments have shown the temperature should be increased to melt the same quantity of solid phase larger by one order of magnitude than maximum possible temperature fluctuation within the accuracy of holding temperature. Thus, the separation of the dendritic branches is nothing else than the dissolution of the most enriched impurity sections during approach to an equilibrium state of the system. The dissolution is not associated with temperature increase but is caused by solidification of very thin specimen (it should be noted that the separation of the branches is observed in any specimen provided temperature increase (Jackson et al.).

The revealing of the features of growth and subsequent dissolution of the dendrites in thin layer was the aim of the further experiments.

Specimens with inhomogeneously distributed impurity

The vessel walls are unpenetrable barrier for impurity during crystal growth. The insoluble inclusions in the melt may be such a barrier, too. Therefore one can suggest that the dissolution of the solid phase after crystallization is to be observed in region with insoluble inclusions similarly to one in very thin sample.

This assumption was verified by using cyclohexanol-water and succinonitrile-salol systems in the specimen having 0.5 mm thickness and containing lycopodium particles. An usual dendritic growth was observed with no particles and was not accompanied by a separation of the dendritic branches during isothermal holding between T_l and T_s. I.e. from the point of view of the transformation mechanism of the dendritic branches such a specimen may be believed to be a bulk one. At the same time, for the specimen of such thickness a stir of the melt due to convection will be important scarcely. For alloys with high concentration of the second component the dendrite is split by no particle proper but by enriched impurity region around a particle. The dendrites formed as a result of splitting grow independently. As seen in Figure 5 besides splitting on solid particles one of the dendritic branches is observed to be separated and then to grow independently. That is, under isothermal conditions, at first, the dendrite grows and then a dendritic section grown in the most enriched-impurity region is dissolved in the spacing between two particles, Figure 5. At the same time another dendritic sections grow continuously without any peculiarities.

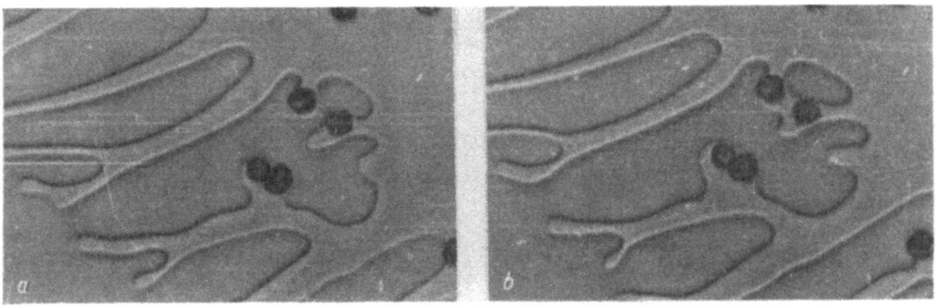

Fig. 5. Interaction between dendrites and lycopodium particles in cyclohexanol-water system

Thus, if there is a barrier unpenetrable for impurity before the growing dendrite then an advance of the dendrite in the impurity-enriched region adjacent to a barrier becomes nonmomotonous: at first, crystallization runs and then a part of crystallized material is dissolved. As follows from the experiments an insoluble inclusion or a vessel wall may be such a barrier. In this case a dissolution effect under isothermal conditions is observed in very thin samples where there is no stir practically in the melt with the result that the large gradient of impurity concentration may arise during growth near a barrier.

If the last statement is true then growth-dissolution effect is to be observed at any case with the large gradient of impurity concentration in the melt independently on that either it is arisen as a result of impurity barrier screening or by some another way.

The crystal growth was studied under conditions of the large concentration gradient within a series of experiments. The gradient was created by introducing of camphene and salol drops into succinonitrile melt. The results of these experiments are as follows. After finishing growth and subsequent isothermal holding the dendrite or the cell having been crystallized in concentration-inhomogeneity region is partially or completely dissolved. When camphene inclusions are present in succinonitrile a sharp boundary is formed between these two slightly soluble substances and it is observed the dissolution of crystal which surface continues to grow. For the dendritic structure forming in the same specimen with temperature decrease a partial dissolution of some bulk having been crystallized is also typical. In this case not only the necks of

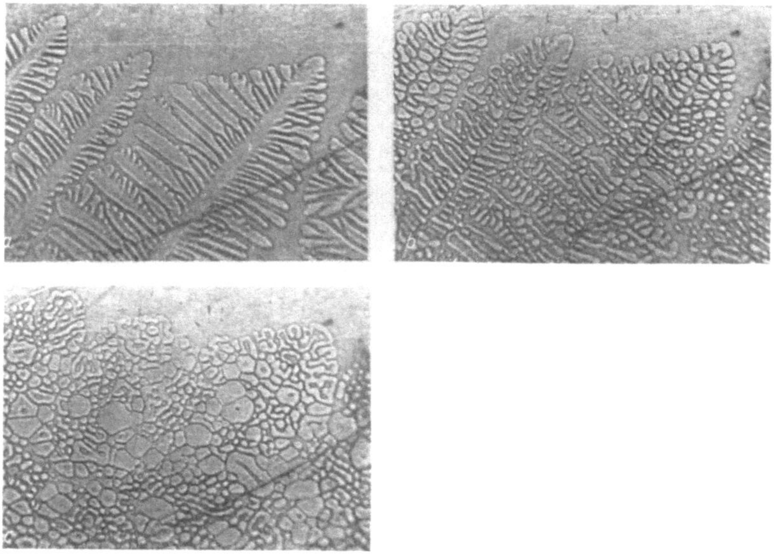

Fig. 6. Dendritic structure transformation in region with increased concentration of camphene: a) 0, b) 10, c) 60 s

the dendritic branches are dissolved but the whole branches. As a result of branches melting and due to appearance of large number of branch fragments a fine-grain structure is formed (Fig. 6).

After a few crystallization-melting cycles the initial concentration gradient decreases owing to which the dissolution becomes weak. That is, depending on a value of the concentration gradient at constant temperature, strong dissolution of the branches, partial dissolution of the necks and, at last, no dissolution effect may be observed. With increasing number of the experiments such a sequence is observed in the same specimen.

Displacement of the crystallization front (or separate dendrite) as a function of time is shown in Figure 7 schematically. With increasing number of the experiments, i.e. as inclusion begins to dissolve more and more the curve 1 changes into curve 2 and then into 3. The above picture is related to definite temperature and to definite rate of the crystallization front far from inclusion. Depending on the conditions of experiment (sample thickness, a size of introduced drops) a certain minimum rate exists below which the process is characterized by curve 3, i.e. without dissolution. This rate corresponds to plane or slightly distorted front.

As follows from the above experiments a part of crystallized material may be dissolved under isothermal conditions in concentration inhomogeneity region in the melt. It means that under definite conditions (growth rate, concentration gradient) the quantity of solid phase is formed more than prescribed by the phase diagram. The appearance of liquid phase after crystallization has been observed before in (GRECHNY) and attributed to "diffusionless" crystal growth in phase diagram region below the line C_0 (Fig. 8). The experiments discussed above seems to confirm directly "diffusionless" crystallization in two-phase region between T_1 and T_s. It would be emphasized that the dissolution of the solid phase resulted from preceding "diffusionless" component of growth rate may be revealed at constant temperature provided the concentration gradient exists in the melt.

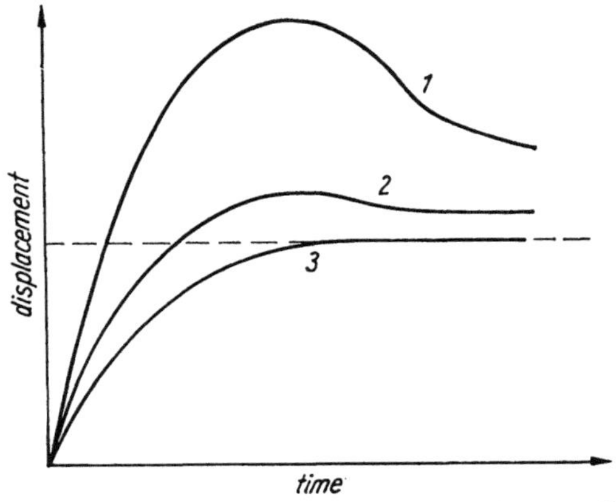

Fig. 7. Scheme of front displacement dependence on time

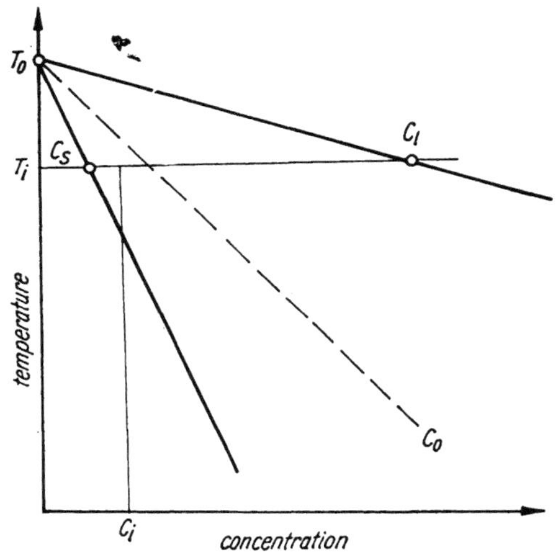

Fig. 8. Schematic phase diagram of two-component alloy

Let us consider crystallization of the melt having the composition C_i and the temperature T_i. According to (KAMENETSKAYA) growth rate, v, includes two components determined by processes of rearrangement, v_r, ("diffusionless" component) and by exchange, v_e, (determined by diffusion). At gradual approach of the crystallized material to an equilibrium composition a remaining liquid composition varies from C_i to C_l. That is, the last part of the crystallized material will have the component v_e only and the dependence of solidified volume on time will be described by curve 1 in Figure 7. If the crystal growing from the melt of the concentration C_i encounters the impurity-enriched region, $\Delta C \geqq C_l - C_0$, it enters into the melt of the concentration C_l without passing through the diffusion-limited growth zone. In as much as the composition of the growing crystal differs from the equilibrium one (due to growth in the $C < C_0$ region) it begins to dissolve up to establishing of local equilibrium. It may be realized if the concentration gradient, ΔC, satisfies the condition

$$\frac{\Delta C}{v/D} > C_l - C_0 \,. \tag{3}$$

Fulfilment of (3) means that at the diffusion length of the growing crystal, v/D, concentration varies from the value corresponding to diffusionless crystallization region ($C < C_0$) to the value above the liquidus temperature ($C \geqq C_l$). The formula (3) seems to give overestimated value of ΔC. In fact, the abovementioned concentration difference should be formed at distance less than v/D because crystal growth is determined by both v_e and v_r.

It has been shown (JACKSON et al.) that a part of crystallized material is dissolved and the dendritic branches are separated from a steam due to thermal fluctuations. In this case the dendritic necks, i.e. the most enriched-impurity sections, are melted. As follows from the data above, the introduction of local concentration inhomogeneities results in the same effect under isothermal conditions. It allows to consider such inhomogeneities as a useful tool to control the solidification structure.

Conclusions

1. In the homogeneous two-component melt under isothermal conditions the dendritic structure is coarsened without separation of the dendritic branches. The separation of the dendritic branches from a steam is observed in the isothermal specimen with a thickness comparable with the dendritic diameter. The similar picture is observed during dendritic growth near an impurity-unpenetrable barrier (solid inclusion) in a thin specimen. This effect is not connected with the dendritic structure transformation as affected by a surface tension but it results from the dissolution of solid phase during approach to an equilibrium.

2. As a consequence of the dendritic growth under the concentration gradient in two-component melt a part of crystallized material is dissolved during isothermal holding after the end of crystallization. The concentration gradient may be formed in the melt both before the beginning of crystallization and after it due to acculmulation of impurity near a barrier. The dissolution of the crystallized material seems to be resulted from the diffusionless growth in the two-phase region of phase diagram.

References

FLEMINGS, M. C.: Solidification Processing, New York 1974

GRECHNY, J. V.: in: Kristallizatsiya i fazovye perekhody, Minsk, 1962, p. 156

HUANG, S. C., GLICKSMAN, M. E.: Acta Metallurg. **29** (1981) 717

JACKSON, K. A., HUNT, J. D., UHLMANN, D. R., SEWARD, T. P.: Trans. Met. Soc. AIME **236** (1966) 147

KAHLWEIT, M.: Scripta Met. **2** (1968) 251

KAMENETSKAYA, D. S.: Ph. D. Thesis, Moscow 1966

KLIJA, M. O.: Kristallografiya **1** (1956) 577

VOORHEES, P. W., GLICKSMAN, M. E.: in: Proc. Symp. Fall. Meet. Met. Soc. AIME (St. Louis 1982), Warrendale, Pa (1983) p. 63

(Received September 7, 1987; accepted September 12, 1987)

Authors' address:

Dr. O. P. FEDOROV, Prof. D. E. OVSIENKO
Institute of Metal Physics, Ukrainian Academy of Sciences
Vernadskogo str. 36
252142 Kiev, USSR

Cryst. Res. Technol. | 23 | 1988 | 4 | 498

Book Review

J. Kepler

Vom sechseckigen Schnee (Strena seu de Nive sexangula)

Ins Dt. übertr., eingel. u. mit Anm. vers. von D. Goetz

Akad. Verlagsgesellsch. Geest & Portig K.-G. Leipzig 1987, 60 S., 16 Abb. u. e. Portr. d. Verf., 8,50 M, ISBN 3-321-00021-0

Betrachtungen zur Geschichte der Kristallographie im 17. Jahrhundert wären unvollständig, würde man die Keplersche Schrift „Vom sechseckigen Schnee" außer acht lassen, die nunmehr erneut in einer deutschen Übersetzung einem breiten Leserkreis zugänglich geworden ist. In ihr wird erstmalig berichtet, daß Schnee in der Form von regelmäßigen sechseckigen Sternchen fällt, daß es aber auch Schneegraupeln gibt. Die Frage nach der Ursache für diese hexagonale Symmetrie durchzieht den gesamten Aufsatz, wobei Kepler zum Vergleich auch andere Beispiele aus der Natur, wie Bienenwaben, Granatapfelkerne und Erbsen heranzieht und am Ende auf Blüten von Pflanzen und sogar auf Kristalle verweist. Bei seiner Suche beschreitet Kepler einen damals ungewöhnlichen, aber heute jedem Kristallographen vertrauten Weg: Er benutzt Modelle aus gleichgroßen Kugeln und findet sogar ihre dichteste Packung heraus. Sein Experiment zum Preßverhalten weicher Kugeln kann als erster Ansatz zu einer Geometrie organischer Formen angesehen werden.

Dem Originaltext beigefügte Erläuterungen, die am Ende des Buches aufgelistet sind, erleichtern das Verständnis der Abhandlung in fachlicher und historischer Hinsicht. Eine vom Übersetzer gegebene Einführung eröffnet Einblick in Keplers Prager Schaffensperiode von 1600—1612, in der die Schrift entstanden ist. Sie muß als ein Vorläufer zu Keplers größerem Werk „Weltharmonik" angesehen werden, das 1619 fertiggestellt wurde, und kündet von den Bestrebungen des Autors, die Harmonie der Welt nicht nur im Planetensystem, sondern auch in einzelnen Naturerscheinungen auf der Erde zu suchen und aufzudecken. Die vorliegende Schrift Keplers sollte als eine der ersten wissenschaftlichen Arbeiten über Kristallographie betrachtet werden und nicht nur bei Wissenschaftshistorikern, sondern auch praktisch tätigen Naturwissenschaftlern und allen naturwissenschaftlich interessierten Lesern Anerkennung finden. H.-J. Höbler

Cryst. Res. Technol.	23	1988	4	499–507

G. Zahn, P. Paufler

Wissenschaftsbereich Kristallographie, Sektion Chemie, Karl-Marx-Universität Leipzig

Identification of Predominant Point Defects in Nonstoichiometric $CuInSe_2$ by X-ray Powder Diffraction

Dedicated to Professor Hermann Neels on the occasion of his 75th birthday

Refinement of structural parameter, occupancy factors and temperature factors of three indium rich samples enabled type and concentration of intrinsic point defects to be determined. Vacancies on copper and indium sites (8 and 5%, respectively) and indium atoms on copper sites (5%) were identified. Indium atoms exhibit a large thermal vibration amplitude.

Die Verfeinerung des Strukturparameters, der Besetzungsfaktoren und der Temperaturfaktoren von drei indiumreichen Proben gestattete die Bestimmung von Typ und Konzentration der Eigenfehlstellen. Leerstellen auf Kupfer- und Indium-Plätzen (8 bzw. 5%) und Indiumatome auf Kupfer-Plätzen (5%) konnten nachgewiesen werden. Indiumatome weisen hohe Werte der Wärmeschwingungsamplitude auf.

1. Introduction

The semiconducting compound $CuInSe_2$ is a promising candidate for optoelectronic device production in particular for solar cells (cf., e.g., Coutts et al.). The crystal structure has been first studied by Hahn et al. using X-ray powder method. The compound belongs to the chalcopyrite ($CuFeS_2$ or $E1_1$) structure type (space group $I\bar{4}2d$). Atomic positions in this structure type are as follows (International Tables for Crystallography (1983); see Fig. 1):

Cu	4a	0, 0, 0; 1/2, 0, 1/4; 1/2, 1/2, 1/2; 0, 1/2, 3/4
In	4b	0, 0, 1/2; 1/2, 0, 3/4; 1/2, 1/2, 0; 0, 1/2, 1/4
Se	8d	x, 1/4, 1/8; $\bar{x}$, 3/4, 1/8; 1/4, $\bar{x}$, 7/8; 3/4, x, 7/8;
		$x + 1/2$, 3/4, 5/8; $1/2 - x$, 1/4, 5/8; 3/4, $1/2 - x$, 3/8; 1/4, $x + 1/2$, 3/8 .

The structure parameter x was determined from X-ray diffractometer measurements to 0.235 ± 0.005 by Parkes et al. At 810 °C $CuInSe_2$ undergoes a phase transition from chalcopyrite to zinc blende by an order-disorder transition in the cation positions 4a and 4b (Palatnik, Rogačeva; Binsma et al.). During cooling crystals below the transition temperature are most often broken into domains (Manolikas, Spyridelis).

Several authors have given results that indicate a range of homogeneity of this compound (Palatnik et al.; Konešova et al.; Chippaux, Deschanvres; Bachmann et al.; Boehnke; Boehnke, Kühn). Intrinsic defects that necessarily arise have an important impact upon electrical transport properties of undoped $CuInSe_2$. Electrical conductivity can easily be changed from n-type to p-type and vice versa by annealing under suitable conditions (Tell et al.; Gombia et al.). The chemistry of structural defects in this compound has recently been surveyed by von Bardeleben (1986).

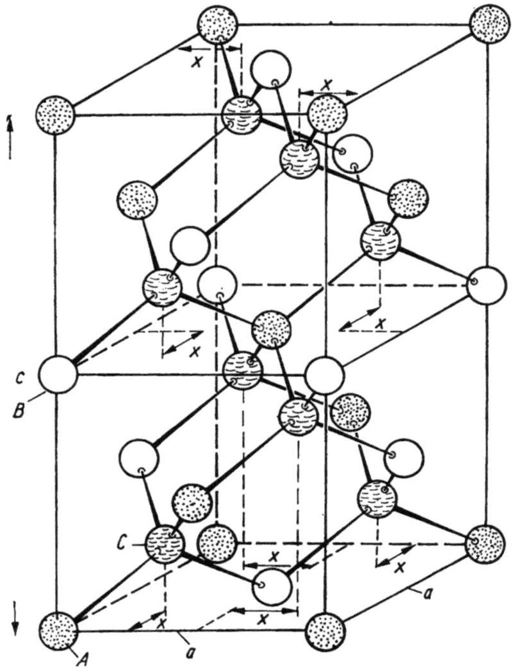

Fig. 1. Unit cell of the chalcopyrite type

No direct observation of intrinsic point defects in $CuInSe_2$ has hitherto been published. To establish specific defect models the study of various physical properties has been used (cf., e.g., MASSÉ, REDJAI; RINĆON, GONZÁLEZ; GONZÁLEZ et al.)

The aim of the present communication is to report on a direct determination of intrinsic point defects by X-ray diffraction. Powder was used to avoid extinction correction. Atomic positions mentioned above were taken as a starting point for the refinement of structural parameter x as well as of site occupancy, temperature and scale factors.

2. Experimental

Bridgman crystals grown by OPPERMANN from ingots prepared by direct synthesis of the elements were crushed and passed through 30 μm mesh screen. The final samples were obtained by sedimentation of an alcoholic suspension of the powder.

Intensity measurements were made with the aid of a two-circle powder diffractometer HZG4 (Freiberger Präzisionsmechanik). Three samples have been studied at 300 K, one sample has been investigated at three different temperatures (80 K, 190 K, and 300 K).

The $\omega - 2\theta$-scan method, filtered CuKα radiation and 1.5° divergence and receiving slits were used in all cases. Time constant and scanning speed were optimized with respect to the slit system. Intensity measurements were made in intervals and over a 2θ range given in Table 1. Dead time corrections were applied automatically during data collection. Lattice constants were determined by adding silicon powder (grain size 15 μm) as internal standard to the specimens. For each sample data sets were collected without removing it from the diffractometer. They were obtained with the X-ray tube operated at 40 kV and 30 mA using step counting times of 20 ... 30 s. Intensity variation due to the variation of sample area stroken by the primary beam between starting and final diffraction angles amounted to 1.5% and was taken into account.

Table 1
Sample and X-ray data collection details

Specimen No	(1)	(2)	(3)
mass density (10^3 kg m^{-3})	5.726(7)	5.689(7)	5.727(3)
chemical composition (%)			
Cu	23.73 (24.4)*)	23.19	poor
In	25.62 (25.6)*)	25.79	rich
Se	50.64 (50.0)*)	51.02	
crystallite size (μm)	15	25	20
absorption coefficient (cm^{-1})	78.6	78.4	78
intervals of measurement (°)	14 ... 144	14 ... 144	22 ... 92
2θ scan range (°)	0.05	0.05	0.1
number of steps in full pattern	2600	2600	700
step counting time (s)	20	20	30
number of unique reflections	45	45	24
number of reflections with intensity $I_{obs} > 3\delta$	38	42	16
number of structure amplitudes F_{hkl}	112	112	50
temperature of measurement (K)	300	300	80, 190, 300

*) Values without parantheses are results by Köhler and in parentheses by Boehnke.

Chemical composition of the samples was taken from the work of Boehnke and Köhler. They used electron microprobe analysis (Boehnke) or coulometry for copper and indium and gravimetry for selenium (Köhler). Mass density was measured by Klöss with the aid of the buoyant force. Sample and X-ray data collection details are summarized in Table 1.

3. Data reduction

The relative integrated peak intensities which were the initial data in the refinement have been evaluated by summing-up the counting rates in the peak area taking background correction into account. Simultaneously, corrections with regard to polarisation, Lorentz and geometry factors were performed.

In principle, no peak profile analysis has been carried out. Only for deconvolution of partially overlapping peaks some few standard reflections were analysed. Peak profiles were best fitted to Cauchy shape functions of the type

$$N_i = N_{max} \left\{ \left[1 + \left(\frac{2\theta_i - 2\theta_{\alpha 1}}{\frac{1}{2} b} \right)^n \right]^{-1} + \frac{1}{2} \left[1 + \left(\frac{2\theta_i - 2\theta_{\alpha 2}}{\frac{1}{2} b} \right)^n \right]^{-1} \right\} \quad (1)$$

where N_i is the background corrected counting rate on the double Bragg angle $2\theta_i$, N_{max} the peak height, $2\theta_{\alpha 1}$, $2\theta_{\alpha 2}$ the peak position, b the peak width and n is an open parameter. For width function a quadratic ansatz with three coefficients was used.

For the further procedure of deconvolution and for the actual refinement the program written by Mortier has been applied. It was modified with respect to crystal structure peculiarities and recording conditions (Zahn 1987). The deconvolution based on Fourier analysis of standard reflections and on a fit of these patterns to the separable overlapping peak profiles.

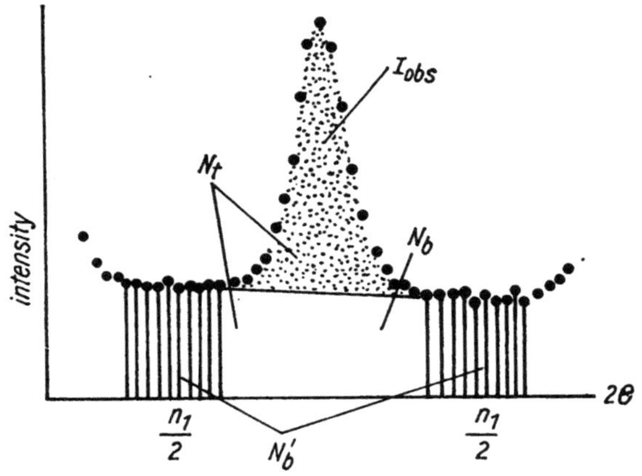

Fig. 2. Peak with background

The standard deviations of the relative integrated intensities have been calculated as follows (cf. Fig. 2):

$$\sigma^2_{I_{obs}} = \frac{1}{(PLG)^2}\left[N_t + \left(\frac{n_2}{n_1}\right)^2 N_b\right] \quad (2)$$

where P, L, G are polarisation, Lorentz and geometry factors, N_t the sum total of counting rates in the peak area, N_b the background in the peak area, N_b' the sum of counting rates used for background determination, n_2, n_1 the number of measuring points in the peak area and for background determination, respectively. This expression takes the precision of background determination into account. The resulting intensities and deviations were the input data in the least-squares refinement.

The structure amplitude F_{hkl} was used in the form:

$$F_{hkl} = \sum_j g_j f_j \exp\left[-2\pi i(hx_j + ky_j + lz_j)\right] \exp\left[-B_j\left(\frac{\sin\theta}{\lambda}\right)^2\right], \quad (3)$$

where g_j is the occupancy factor of site j, f_j the tabulated complex atomic scattering factor, h, k, l Miller indices of the reflection, x_j, y_j, z_j coordinates of site j, B_j the isotropic Debye-Waller factor, θ the Bragg angle and λ the X-ray wave length.

4. Results

Numerical results of refinement and lattice constants a, c are given in Table 2 and 3, respectively. Definitions of included reliability factors R see at bottom of the Table 2. w is the weighting factor, x is the structure parameter and s the scale factor. Subscripts 'calc' and 'obs' mean 'calculated' and 'observed', respectively. If mass density and chemical composition values were taken for first proposals of site occupancies, refinement of specimen (1) yielded reasonable results whereas for specimens (2) and (3) additional constraints had to be inserted. These constraints included permanent agreement between occupancy factors, mass density and chemical composition within limits of error.

Since some other phases had been detected in the diffraction pattern of copper rich material, only indium rich specimens have been investigated. The binary metal

Table 2
Site occupancy factors g, crystal structure parameter x, isotropic Debye-Waller factors B (in 10^{-20} m^2) and R values (e.s.d. in parentheses)

Specimen No		(1)	(2)	(3) 80 K	(3) 190 K	(3) 300 K
Cu_{Cu}	g	0.923(22)	0.938(27)	0.944(23)	0.945(26)	0.947(28)
	B	1.65(28)	1.26(55)	0.49(51)	0.86(57)	1.33(67)
In_{In}	g	0.946(12)	0.976(18)	0.950(15)	0.938(18)	0.947(18)
	B	2.87(24)	3.01(44)	1.19(38)	1.61(57)	2.28(65)
Se_{Se}	x	0.2268(7)	0.2281(14)	0.2233(17)	0.2226(16)	0.2227(16)
	g	1.017(15)	0.975(20)	1.002(15)	0.996(18)	0.999(17)
	B	1.23(19)	0.77(27)	1.25(37)	1.44(45)	1.46(55)
In_{Cu}	g	0.054(11)	0.053(16)	0.054(13)	0.052(14)	0.055(15)
	B	2.87	3.01	1.19	1.61	2.28
In_i	g	0.001(8)	0.001(10)	—.002(5)	—	—
	B	2.87	3.01	1.19	—	—
R_{wI}		0.0388	0.0498	0.0368	0.0376	0.0407
R_I		0.0391	0.0825	0.0515	0.0471	0.0654
R_{wF}		0.0389	0.0978	0.0407	0.0381	0.0468
R_F		0.0478	0.0978	0.0509	0.0548	0.0595

$R_{wI} = [\sum w(I_{calc} - sI_{obs})^2/\sum w(sI_{obs})^2]^{1/2}$
$R_I = \sum |I_{calc} - sI_{obs}|/\sum (sI_{obs})$
$R_{wF} = [\sum w^{1/2} (|F_{calc}| - F_{obs})^2/\sum w^{1/2}F^2_{obs}]^{1/2}$
$R_F = \sum ||F_{calc}| - F_{obs}|/\sum F_{obs}$
$w = \sigma^{-2}_{I_{obs}}$

selenides present were structure analogues of $CuInSe_2$, thus overlapping peaks prevented crystal structure refinement.

R values of 4 ... 5% compare favourably with other powder refinement results.

5. Discussion

The unit cell size lies within the range of results reported by Hahn et al., Palatnik et al., Parkes et al., and Spiess et al. The structural parameter x is somewhat smaller than that of Parkes et al. and comparable to the result of Spiess et al. (cf. Table 3). Since the former authors did not give the chemical composition of their samples we suppose an influence of nonstoichiometry. Moreover, they used only 10 reflections for refinement and assumed ideal occupancy. Values of x less than 1/4 mean a displacement of selenium atoms from the 'ideal' position $x = 1/4$ towards the copper atoms. The variation of x within the range of homogeneity shows no clear correlation to the occupancy factors. The results of specimen (3), however, are less reliable because the (101) reflection could not be recorded due to the geometrical arrangement of cryostat windows. Just the low angle reflections are especially important for the determination of x (cf. Parkes et al.; Jaffe, Zunger; Zahn (1982)).

Several attempts have been made to correlate x with different physical properties. Most of them were, however, restricted to deviations from the regularity of the Se_4 tetrahedron as put forward by Abrahams and Bernstein (1973). Recently, Paufler and Rotsch have distinguished 18 classes of geometrical relations between the axial

Table 3
Lattice constants a, c, bond lengths (in 10^{-10} m), axial ratio c/a, and structure parameter x of some chalcopyrite type compounds

ABC_2	a	c	c/a	x	A—C	B—C	ref.
$CuGaSe_2$	5.614(1)	11.03(1)	1.965(2)	0.250(1)	2.417(3)		SPIESS et al.
	5.5963(1)	11.0036(2)	1.96623(5)	0.2431(2)	2.3875(7)		SCHNEIDER et. al.
$CuAlSe_2$	5.602(2)	10.944(5)	1.954(1)	0.269(5)	2.471(17)		SPIESS et al.
$AgInSe_2$	6.109(1)	11.717(5)	1.919(1)	0.250(1)		2.610(4)	SPIESS et al.
	6.1038(15)	11.7118(47)	1.91877(90)	0.25845(16)		2.5779(7)	BENOIT et al.
$CuInSe_2$	5.784(1)	11.616(5)	2.008(1)	0.224(3)	2.424(9)	2.598(11)	SPIESS et al.
	5.782(1)	11.620(1)	2.0097(4)	0.235(5)	2.459(16)	2.559(17)	PARKES et al.
Specim. (1)	5.784(1)	11.614(4)	2.008(1)	0.2268(7)	2.433(2)	2.588(3)	this work
Specim. (2)	5.781(1)	11.609(3)	2.008(1)	0.2281(14)	2.436(4)	2.582(5)	this work
Specim. (3)	5.783(1)	11.611(6)	2.008(2)	0.2227(16)	2.420(5)	2.602(6)	this work

ratio c/a and x. $CuInSe_2$ belongs to the class the structure of which is governed by the equality of nearest interatomic distances between copper and indium atoms parallel and oblique to the basal plane whatever the axial ratio would be. Note that this is equivalent to the statement that interaction of non-next-nearest atoms considerably influences structural geometry. The present results confirm the attractive interaction between selenium and copper already known from previous measurement of x.

The site occupancy factors of Table 2 enable the predominant types of intrinsic point defects to be identified. Thus, in indium rich $CuInSe_2$ the following defects were detected:

(i) Vacancies on copper and indium sites (V_{Cu}, V_{In}) at maximum of the order of $(8 \pm 2)\%$ and $(5 \pm 1)\%$, respectively,
(ii) indium atoms on copper sites (In_{Cu}) of the order of $(5 \pm 1)\%$.

The concentration of interstitials of indium atoms In_i was beyond the limit of accuracy. Standard deviations given in Table 2 lead to the conclusion that all selenium positions should be considered occupied, although some selenium vacancies could not completely be ruled out in specimen (2).

For both the ionic and the covalent bonding model, V_{Cu} and V_{In} should introduce acceptor levels (NEUMANN; BARDELEBEN (1984); MASSÉ, REDJAI) and In_{Cu} should appear as donor (NEUMANN). This could explain why often p-conductivity is observed in copper rich samples, whereas some compensation may occur in indium rich material. Transport measurements made by NEUMANN et al. revealed the presence of

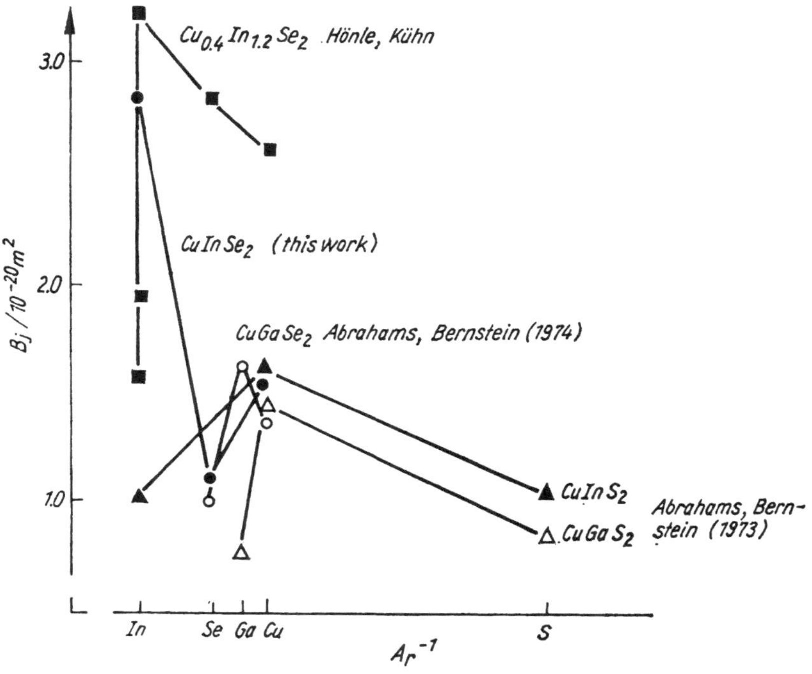

Fig. 3. Isotropic temperature factors of atoms for analogous chalcopyrite type compounds (For the single crystal studies quoted isotropic Debye-Waller factors were calculated from $B = 1/3 \sum_{i=1}^{3} b_{ii}$.)

acceptor and donor levels in bulk material which they attributed to V_{Se}, V_{In}, Se_i, and Fe_{In}. Specimens (1) and (3) of the present work were investigated by NOWAK. He interpreted the five levels found as V_{Cu}, V_{In}, In_{Cu}, Cu_{In} and In_i.

The temperature factor $B = 8\pi^2\overline{u}^2$ as a function of the reciprocal atomic mass is shown in Figure 3. $\overline{u}^2$ denotes the mean-square deviation of an atom from its position coordinate. Note that the thermal vibration amplitudes of indium atoms are considerably larger than those of selenium and copper and reach large absolute values. Results of the other chalcopyrite type compounds and of $Cu_{0.4}In_{1.2}Se_2$ are given for comparison. The results of HÖNLE and KÜHN with latter compound crystallizing in a similar structure support the unusual vibration amplitude of indium found in the present refinement.

Furthermore, Table 3 contains crystal structure details of chalcopyrite type compounds which consist of at least two of the elements copper, indium and selenium.

To sum up, the present results give for the first time absolute figures of intrinsic defect concentration in nonstoichiometric $CuInSe_2$. The defect types found are compatible with findings of indirect measurements.

The authors are indebted to Mr. D. Oppermann for supply of the specimens and to Prof. G. Kühn for helpful discussions. One of us (G. Z.) wants to express his gratitude to Prof. K. Kleinstück for the kind hospitality during a post-gradual study when the low-temperature measurements were carried out at the Technical University Dresden.

References

ABRAHAMS, S. C., BERNSTEIN, J. L.: J. Chem. Phys. **59** (1973) 5415
ABRAHAMS, S. C., BERNSTEIN, J. L.: J. Chem. Phys. **61** (1974) 1140
BACHMANN, K. J., FEARHEILY, M., SINGH, Y. H., TRAN, N.: Appl. Phys. Lett. **44** (1984) 407
BARDELEBEN, H. J. VON: Solar Cells **16** (1986) 381
BARDELEBEN, H. J. VON: J. Appl. Phys. **56** (1984) 321
BENOIT, P., CHARPIN, P., LESUEUR, R., DJAGA-MARIADASSOU, C.: Jap. J. Appl. Physics Suppl. **19-3** (1980) 85
BINSMA, J. J. M., GILING, L. J., BLOEM, J.: phys. stat. sol. (a) **63** (1981) 595
BOEHNKE, U.-C.: Thesis, Karl-Marx-Universität Leipzig 1985
BOEHNKE, U.-C., KÜHN, G.: J. Mater. Science **22** (1987) 1635
CHIPPAUX, D., DESCHANVRES, A.: J. Solid State Chemistry **45** (1982) 200
COUTTS, T. J., KAZMERSKI, L. L., WAGNER, S. (eds): Copper Indium Diselenide for Photovoltaic Applications, Amsterdam 1986
GOMBIA, E., LECCABUE, F., SALVIATI, G., SEURET, D.: J. Crystal Growth **65** (1983) 270
GONZÁLEZ, J., RINĆON, C., REDONDO, A., NEGRETE, P.: Jap. J. Appl. Physics Suppl. **19-3** (1980) 29
HAHN, H., FRANK, G., KLINGLER, W., MEYER, A.-D., STÖRGER, G.: Z. Anorg. Allg. Chemie **271** (1953) 153
HÖNLE, W., KÜHN, G.: (to be published)
International Tables for Crystallography Vol. A, ed. by Th. HAHN, Dordrecht 1983
JAFFE, J. E., ZUNGER, A.: Phys. Rev. B **28** (1983) 5822
KLÖSS, G.: Thesis, Karl-Marx-Universität Leipzig 1987
KÖHLER, A.: Diploma thesis, Karl-Marx-Universität, Sektion Chemie, Leipzig 1983
KONEŠOVA, T. I., BABICYNA, A. A., KALLINIKOV, V. T.: Izv. AN SSSR, Neorg. Mater. **18** (1982) 1483
MANOLIKAS, C., SPYRIDELIS, J.: Mat. Res. Bull. **16** (1981) 501
MASSÉ, G., REDJAI, E.: J. Phys. Chem. Sol. **47** (1986) 99
MORTIER, W. J.: Programs for Data Sampling, Structure Refinement, and Statistical Evaluation of Structure based on Powder Data, Katholieke Universiteit Leuven, Belgium, 1979

NEUMANN, H.: Cryst. Res. Technol. **18** (1983) 483
NEUMANN, H., NOWAK, E., KÜHN, G.: Cryst. Res. Technol. **16** (1981) 1369
NOWAK, E.: Thesis, Karl-Marx-Universität Leipzig 1983
OPPERMANN, D.: Diploma thesis, Karl-Marx-Universität, Sektion Chemie, Leipzig 1982
PALATNIK, L. S., KOMNIK, J. F., ROGAČEVA, E. I.: Ukr. Fiz. Žurn. **9** (1964) 862
PALATNIK, L. S., ROGAČEVA, E. I.: Dokl. AN SSSR **174** (1967) 80
PARKES, J., TOMLINSON, R. D., HAMPSHIRE, M. J.: J. Appl. Cryst. **6** (1973) 414
PAUFLER, P., ROTSCH, P.: Z. Krist. **172** (1985) 183
RINĆON, C., GONZÁLEZ, J.: phys. stat. sol. (b) **110** (1982) K171
SCHNEIDER, J., RÄUBER, A., BRANDT, G.: J. Phys. Chem. Sol. **34** (1973) 443
SPIESS, H. W., HAEBERLEN, U., BRANDT, G., RÄUBER, A., SCHNEIDER, J.: phys. stat. sol. (b) **62** (1974) 183
TELL, B., SHAY, J. L., KASPER, H. M.: J. Appl. Physics **43** (1972) 2469
ZAHN, G.: Diploma thesis, Karl-Marx-Universität, Sektion Chemie, Leipzig 1982
ZAHN, G.: Thesis, Karl-Marx-Universität Leipzig 1987

(Received, accepted September 11, 1987)

Authors' addresses:
Dr. rer. nat. Gernot ZAHN (a), Prof. Dr. sc. nat. Peter PAUFLER (b)
Karl-Marx-Universität Leipzig, Sektion Chemie
Wissenschaftsbereich Anorganische Chemie (a) and
Wissenschaftsbereich Kristallographie (b)
Talstr. 35
Leipzig
DDR-7010

Book Review

G. MÜLLER

Über die Entstehung von Inhomogenitäten in Halbleiterkristallen bei der Herstellung aus Schmelzen

Selisch Fachbuch-Verlag, 1986, 178 Seiten, 102 Abbildungen, 2 Tabellen, Preis DM 45,80, ISBN: 3-88813-000-X

Das vorliegende Buch beinhaltet den Abdruck einer Habilitationsschrift aus dem Jahre 1986. Gleichzeitig wird damit eine neue naturwissenschaftlich-technische Reihe eröffnet.

Die Abhandlung faßt den internationalen Stand und zahlreiche Experimente des Autors zu einem zentralen Kristallzüchtungsproblem zusammen — züchtungsbedingte chemische Inhomogenitäten in Halbleiterkristallen. Die Teile I und II leiten in die Problematik ein und machen den Leser mit den theoretischen Grundlagen des Kristallwachstums, insbesondere der Hydrodynamik in Schmelzvolumina bekannt. Dies geschieht konzentriert und praxisbezogen, unter Verzicht auf umfangreichen mathematischen Apparat. Die Probleme der Segregation (makroskopischen Inhomogenität) insbesondere bei konvektiver Durchmischung der Schmelze werden in Teil III abgehandelt. Äußerst instruktiv sind die Abbildungen berechneter und simulierter Konvektionsbilder in züchtungsähnlichen Containern. Die Ursachen der Streifenbildung in Halbleiterkristallen (mikroskopische Inhomogenitäten) werden im Teil IV allgemein dargestellt und im Teil V auf die instationäre Auftriebskonvektion zurückgeführt. Die Temperatur-Zeit-Experimente in verschiedenen Schmelzanordnungen mit typischen Kristallzüchtungsgeometrien belegen anschaulich diesen Zusammenhang. Der letzte Teil des Buches zeigt Wege zur Reduzierung und Vermeidung konvektionsbedingter Mikroinhomogenitäten, vorrangig unter Zuhilfenahme stationärer Magnetfelder und Zentrifugeanordnungen.

Insgesamt stellt diese Arbeit einen sehr guten Überdlick dar. Einprägsam werden theoretische Voraussagen mit experimentellen Ergebnissen zu den einzelnen Züchtungsanordnungen (Czochralski-, vertikale und horizontale Bridgman-, Zonenkristallisation) verkoppelt, wie wohl überhaupt der besondere Wert des Buches in der Praxisbezogenheit und seiner klar verständlichen Beweisführung liegt. Es ist eine ausgezeichnete Grundlage für den Kristallzüchter und weitere Spezialisten auf den Gebieten der Materialwissenschaften.

P. RUDOLPH

Cryst. Res. Technol. | 23 | 1988 | 4 | 509–515

G. Zahn, B. A. Merisov, G. Klöss, G. Ya. Khadshaj, S. Müller, O. A. Gavrenko, P. Paufler

Wissenschaftsbereich Kristallographie, Sektion Chemie, Karl-Marx-Universität Leipzig, and Kharkov State University, USSR

On the Change of Low-temperature Thermal Conductivity of GaP after Bending

The thermal conduction of a sulphur-doped single crystal plastically deformed by bending at 963 K was measured between 2 and 50 K. The thermal resistivity W at temperatures > 14 K is practically independent on deformation. At lower temperatures W is found to be proportional to AT^{-3} before and to $AT^{-3} + BT^{-2}$ after deformation. A is due to boundary scattering, B due to dislocations.

Die Wärmeleitung eines schwefeldotierten Einkristalls wurde zwischen 2 und 50 K vermessen, der bei 963 K durch Biegung plastisch verformt worden ist. Der Wärmewiderstand W ist bei Temperaturen > 14 K praktisch unabhängig von der Verformung. Bei tieferen Temperaturen erweist sich W als proportional zu AT^{-3} vor und zu $AT^{-3} + BT^{-2}$ nach der Verformung. A geht auf Grenzflächenstreuung, B auf Versetzungen zurück.

1. Introduction

Plastic deformation has considerable impact upon physical properties of semiconductors (cf., e. g. Labusch, Schröter). Although III–V compounds like GaP have become materials attractive for optoelectronic device production only a few studies are devoted to their deformation properties (cf. e.g. Davidson et al.; Gottschalk et al.; Henschel et al.; Retzlaff). No attention has hitherto been paid to the influence of deformation on the thermal conductivity. Muždaba et al. measured the thermal conductivity of as-grown GaP at low temperatures. They did not, however, analyse the scattering mechanisms.

GaP belongs to the zinc blende structure type (space group F$\bar{4}$3m, lattice parameter 0.54505 nm (Landolt-Börnstein)). It exhibits a Debye temperature θ_D of 456 K (Deus et al.). The energy gap between valence and conduction bands of electron states is large (2.35 eV).

The aim of the present paper is to present and analyse results on the change of the thermal conductivity of a GaP single crystal due to plastic bending. Measurements were carried out at low temperatures to avoid complexities arising from simultaneous heat conduction by phonons, electrons, holes, and photons.

2. Experimental

A sulphur-doped liquid-encapsulated Czochralski n-GaP crystal was cut by a diamond saw. The geometry of the specimen is shown in Figure 1. The carrier concentration $N_D - N_A$ was $3.6 \cdot 10^{17}$ cm^{-3}, carrier mobility amounted to 128 cm^2/Vs.

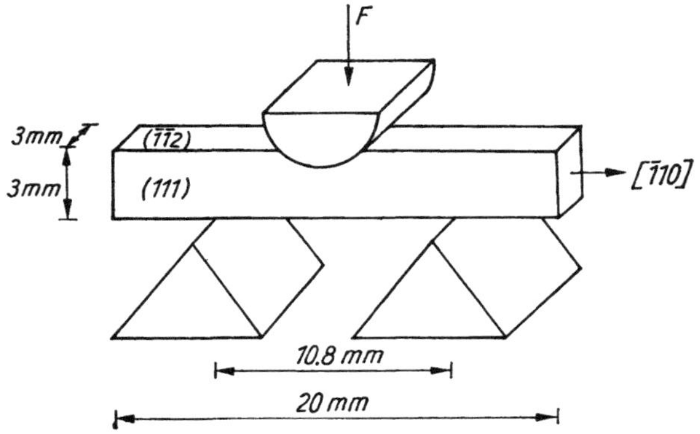

Fig. 1. Geometry of specimen bent

In order to determine the etch pit density prior to deformation, an H_3PO_4 etchant was used at 180—190 °C; the etching time was 3—5 minutes. Etch pit density amounted to $1.5 \cdot 10^5$ cm^{-2}.

Average dislocation density after deformation was calculated after NYE from the radius of curvature R according to

$$N_d^a = \frac{1}{R \cdot b}$$

where R was determined by optical microscopy as described by PAUFLER et al. (b Burgers vector modulus).

The etch method was not applied on the $(\bar{1}\bar{1}2)$ face because of its poor resolution and with respect to the fact that the dislocation density varied throughout the specimen.

The specimen was deformed in a N_2/H_2 atmosphere by means of a three-point bending jig attached to a Polanyi machine. Graphite knife-edges were used (radii 0.5 mm and 10 mm for the outer and inner knifes, respectively). The distance of the outer knife-edges l_k was 10.8 mm. Force F and deflection d of the central knife were measured. The cross-head speed was kept constant at 6 mm/h. The strain rate for d = 50 μm was about $2.3 \cdot 10^{-4}$ s^{-1}. Details of the deformation equipment are described elsewhere (RETZLAFF; MÜLLER). The deformation temperature measured by means of thermocouples was 963 K. Absolute values of the specimen temperature were determined to within ±10 K. Mechanical parameters as taken from MÜLLER are given in Table 1. No upper yield point has been observed.

Mass density was measured prior to and after deformation with the aid of buoyant force. Details of the procedure are given by KLÖSS.

Table 1
Deformation parameters

σ_P MPa	τ_P MPa	σ_{max} MPa	d μm	ε_P %	N_d^b cm^{-2}	N_d^a cm^{-2}
96	13.6	106	424	3.7	$(1.5 \dots 2) \cdot 10^5$	$1 \cdot 10^7$

σ_P yield stress at the beginning of plastic deformation
τ_P critical resolved shear stress in the glide system $\{111\}\langle 1\bar{1}0\rangle$
d deflection
ε_P plastic strain at the surface when plasticity sets in
σ_{max} maximum stress at the end of deformation
N_d dislocation density prior to (b) and after (a) deformation

Measurements of thermal conductivity have been carried out at temperatures between 2 and 50 K by a stationary heat flux method, the flow gradient being parallel to [$\bar{1}$10]. The specimen was coupled at one end to a copper refrigerant the temperature of which was indicated by a carbon resistance thermometer TSU-2. At the other end a small heater was attached. Along a side face of the specimen two Au + 0.07 at.% Fe–Cu thermocouples were glued symmetrically to its centre 10 mm apart. The working difference of temperature along the specimen varied between 0.05 and 0.15 K. All adhesive joints were accomplished by the adhesive BF-2 mixed with highly dispersed copper powder.

3. Results and discussion

The temperature dependence of thermal conductivity λ prior to and after deformation is shown in Figure 2 and Table 2. Below 14 K λ decreases considerably after plastic deformation. The absolute values are generally larger than those reported by MUŽDABA et al. which is probably due to different doping. Both curves are similar in shape and exhibit a maximum at about $0.03\theta_D$, which is the normal behaviour for semiconductors. Thermal resistivity W (see Fig. 3) accordingly shows a sharp minimum. At temperatures above the minimum a linear temperature deopendence

$$W = W_r + CT \tag{1}$$

is observed only slightly dependent on deformation. This is typical of the phonon scattering contribution by point defects (ZIMAN) as, e.g., isotopes, vacancies, antisite defects, and impurities which were detected in GaP by several authors with the aid of various methods (cf. KLIMM). W_r is the residual thermal resistivity due to these point defects. Figures of W_r and C are summarized in Table 3. MOGILEVSKIJ and

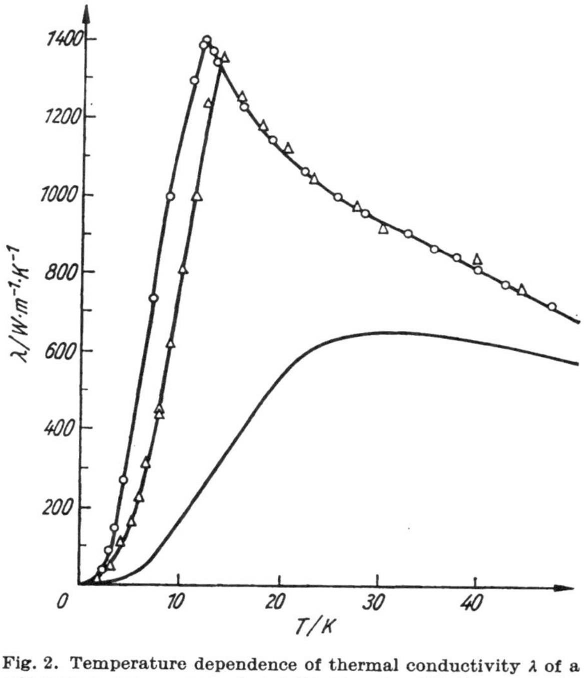

Fig. 2. Temperature dependence of thermal conductivity λ of a GaP single crystal doped with sulfur (○ as-grown, △ bent at 2.7°). Results of MUŽDABA et al. are shown for comparison

Table 2
Thermal conductivity λ as a function of temperature

prior to deformation		after deformation	
T/K	λ/W/m K	T/K	λ/W/m K
2.263	37.6	1.920	13.6
2.974	87.0	2.978	44.3
3.562	144.7	4.220	107.0
4.356	273	5.181	166
7.108	735	6.035	228
8.877	994	6.704	312
11.036	1296	7.965	440
11.901	1387	8.011	452
12.137	1400	9.038	620
12.668	1371	10.222	808
13.172	1347	11.482	994
16.045	1229	12.400	1237
18.750	1145	13.902	1357
22.113	1067	15.708	1252
25.514	1002	17.962	1177
28.226	962	20.282	1124
32.436	911	23.138	1042
35.081	872	27.361	975
37.384	852	30.086	924
39.628	821	39.460	843
42.196	783	43.956	766
47.045	726		

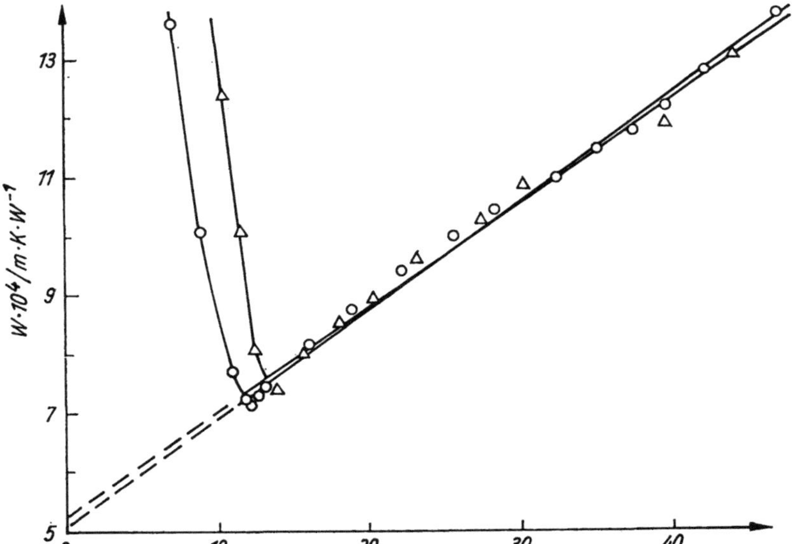

Fig. 3. Temperature dependence of thermal resistivity $W = \lambda^{-1}$, where λ was taken from Figure 2 (○ as-grown, △ after plastic deformation)

Table 3
Constants A, B, C according to equations (1) and (2)

	prior to deformation	after deformation
$A/\mathrm{m \cdot K^4 \cdot W^{-1}}$	0.31 ± 0.01	0.31 ± 0.02
$B/\mathrm{m \cdot K^3 \cdot W^{-1}}$	0	0.102 ± 0.003
$C/10^{-3} \cdot \mathrm{m \cdot W^{-1}}$	1.82 ± 0.04	1.77 ± 0.08

Čudnovskij give $W_r = 4.6 \cdot 10^{-4}$ mK W^{-1} calculated for phonon scattering by gallium isotopes on the basis of their natural distribution. It is of the same order of magnitude as the value given in Table 3.

Mass density measurements D revealed a decrease of specimen density after deformation at 0.17% (before $D_0 = (4.127 \pm 0.001) \cdot 10^3$ kg m^{-3}, after $D_d = (4.120 \pm 0.001) \cdot 10^3$ kg m^{-3}) which should be due to an increase in the concentration of vacancies (Klöss). The slight increase in W_r may be explained by that change.

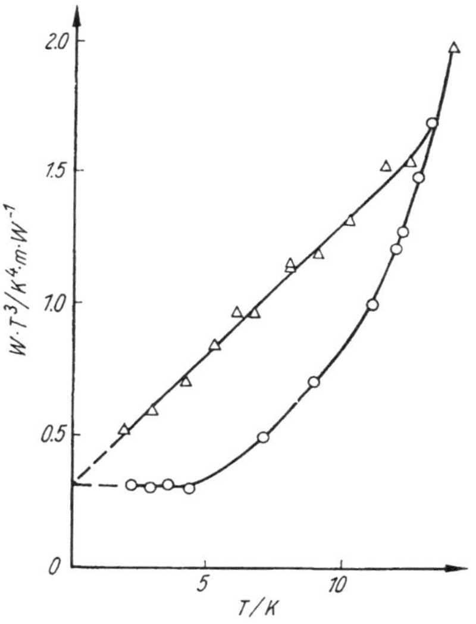

Fig. 4. Separation of contributions due to boundary and dislocation scattering (○ as-grown, △ deformed specimen)

The expected form of W at low temperatures is (cf. Klemens; Mogilevskij, Čudnovskij; Wasserbäch (1978))

$$W(T) = AT^{-3} + BT^{-2}, \tag{2}$$

where AT^{-3} denotes the resistivity of the undeformed specimen limited by surface (boundary) scattering and BT^{-2} represents the thermal resistivity due to dislocations after plastic deformation (Lomer, Rosenberg). To separate both parts it is convenient to plot experimental data as WT^3 versus T (cf. Rosenberg). Results are shown in Figure 4 and Table 3. As is clearly seen in Figure 4, thermal resistivity of the undeformed specimen is governed solely by boundary scattering between 2 and 4.5 K ($B = 0$). The dislocation contribution after deformation is well pronounced within the whole temperature range below the minimum of W. The boundary contribution remained unchanged after deformation (A unchanged, see Table 3). Apparently boundary and dislocation influences are independent, i.e., Matthiessen's rule concerning the additivity of thermal resistances holds.

Theory of boundary scattering leads to (cf., e.g., Berman)

$$A^{-1} = \frac{k_B^4 \cdot J_4(\infty) \cdot d}{2 \cdot \pi^2 \cdot \hbar^3 \cdot v^2}, \tag{3}$$

where k_B is Boltzmann's constant, J_4 the Debye integral, d specimen diameter, $\hbar = h/2\pi$, h Planck's constant, v phonon velocity. From our results we find $v = 6.1 \cdot 10^3$ m s^{-1}. This compares favourably with the longitudinal sound velocity $v = 5.3 \cdot 10^3$ m s^{-1} parallel [110] as derived from the elastic constants $c_{11} = 141$, $c_{12} = 52.5$ and $c_{44} = 70.5$ GPa (Landolt-Börnstein).

The dislocation structure after bending of GaP has been investigated by Paufler et al. They have shown that 60° dislocations predominate. Their average spacing in our specimen is of the order of ≈ 3 µm. Local scattering of phonons by the strain field of dislocations becomes effective and $W - T^{-2}$ is observed. From the results of Tables 1 and 3 a measure of the scattering per dislocation $B/N_d = 1.0 \cdot 10^{-12}$ m^{3}K^{3}W^{-1} can be obtained which is considerably larger than figures reported for niobium and tantalum by Wasserbäch (1985). We can conclude that thermal conductivity is sensitive to dislocations at low temperatures.

The authors express their gratitude to Mrs. B. Wegener for her support of the experimental work. Two of us (G.Z. and G.K.) are indebted to the Low Temperature Laboratory of Kharkov University for the kind hospitality during a postgradual study.

References

Berman, R.: Thermal Conduction in Solids, Oxford 1976

Davidson, S. M., Iqbal, M. Z., Northrop, D. C.: phys. stat. sol. (a) **29** (1975) 571

Deus, P., Schneider, H. A., Voland, U.: Cryst. Res. Techn. **16** (1981) 941

Gottschalk, H., Patzer, G., Alexander, H.: phys. stat. sol. (a) **45** (1978) 207

Henschel, D., Boehnke, O., Paufler, P.: Proc. 6th Int. Symp. on High-Purity Materials in Science and Techn., Dresden 1985, P.-Abstracts, p. 369

Klemens, P. G.: Solid State Physics **7** (1958) 1

Klimm, D.: Thesis, Karl-Marx-Universität Leipzig 1986

Klöss, G.: Thesis, Karl-Marx-Universität Leipzig 1987

Labusch, R., Schröter, W.: in: Dislocations in Solids, Vol. 5 (ed. F. R. N. Nabarro), Amsterdam 1980, pp. 127

Landoldt-Börnstein: Vol. III/18, p. 14, Berlin 1984

Lomer, J. N., Rosenberg, H. M.: Phil. Mag. **4** (1959) 467

Mogilevskij, B. M., Čudnovskij, A. P.: Teploprovodnost' poluprovodnikov, Moskva 1972

Müller, S.: Diploma work, Karl-Marx-Universität, Leipzig 1986

Muzdaba, V. M., Našel'skij, A. Ja., Tamarin, P. V., Šalyt, S. S.: Fiz. Tverd. Tela **10** (9) (1968) 2866

Nye, J. P.: Acta Metallurg. **1** (1953) 153

Paufler, P., Rotsch, P., Wagner, G.: Phil. Mag. **A56** (1987) 533

Retzlaff, U.: Diploma Work, Karl-Marx-Universität Leipzig, Sektion Chemie 1985

Rosenberg, H. M.: Phil. Trans. Roy. Soc. A **247** (1955) 441

Wasserbäch, W.: Phil. Mag. A **38** (1978) 401

WASSERBÄCH, W.: phys. stat. sol. (b) 128 (1985) 453
ZIMAN, J. M.: Electrons and Phonons, Oxford 1960

(Received, accepted September 30, 1987)

Authors' adresses:

Prof. Dr. sc. nat. P. PAUFLER, Dr. G. ZAHN, Dr. G. KLÖSS, Dipl.-Krist. S. MÜLLER
Wissenschaftsbereich Kristallographie, Sektion Chemie der Karl-Marx-Universität
Leipzig Talstraße 35
Leipzig 7010, GDR

Dr. B. A. MERISOV, Dr. G. Ya. KHADSHAJ, O. A. GAVRENKO
Kaf. Niz. Temp., Fiz. Fak., Gos. Univ. imeni Gor'kogo Kharkov
pl. Dzerdžinskogo
Kharkov, USSR

Book Review

P. L. DUBOV, V. A. FRANK-KAMENECKIJ, I. I. ŠAFRANOVSKIJ

Obobščennaja simmetrija

Leningradskij Gosudarstvennij Universitet "A. A. ŽDANOV" 1985. 76 S., 82 Abb., Preis 15 Kop.

Die vorliegende Broschüre enthält eine Ableitung der Klassen von Punktgruppen folgender Symmetrieoperationen: Schwarz-Weiß-Symmetrie, Mehrfarbensymmetrie sowie homologe, konforme, krummlinige und lokale Symmetrie. Die Darstellung ist auf die Bedürfnisse von Studenten der Fachrichtungen Kristallographie, Mineralogie, Petrographie und Geochemie zugeschnitten, eignet sich aber auch als Lektüre für Studenten verwandter Fachgebiete wie Festkörperphysik und -chemie sowie für wissenschaftliche Mitarbeiter der Festkörper- und Werkstofforschung.

Vorausgesetzt wird die Kenntnis der klassischen Symmetrietheorie unter Beachtung einfacher Gruppeneigenschaften einschließlich der zugehörigen Nomenklatur. Dazu ist von den gleichen Autoren im gleichen Verlag 1984 eine in Aufbau und Ausstattung analoge Broschüre mit dem Titel ,,Klassische Symmetrie" erschienen, aber von deren Verfügbarkeit sollte der Zugang zu vorliegender Verallgemeinerung nicht abhängen.

Etwa die Hälfte des Umfangs ist der Antisymmetrie gewidmet. Sie wird auf naivgeometrischem Wege eingeführt, bzgl. strengerer Verfahren findet man Literaturverweise.

Die Einbeziehung der Drehoperation mit Zähligkeit Unendlich ist vorgenommen. Eingehendere Behandlung erfährt die Mehrfarbensymmetrie (3, 4 und 6 Farben) sowie die Vielfachantisymmetrie.

Die andere Hälfte der Studienhilfe ist der nichtisometrischen Symmetrie vorbehalten. Darunter sind die Ähnlichkeitssymmetrie (Homothetie), die konforme, die homologe, die krummlinige und die verallgemeinerte lokale Symmetrie besonders hervorgehoben.

Die Ausführungen sind durchweg von instruktiven Skizzen und von Wiederholungsfragen bzw. Aufgaben begleitet. Sie bilden insgesamt eine sehr begrüßenswerte Einführung in die meist ohne Vorkenntnisse schwer zugängliche Spezialliteratur, und sie geben gleichzeitig eine Übersicht zum Gesamtproblem ohne Vollständigkeit anzustreben. Am Schluß finden sich kurze Hinweise auf Anwendungen. Dieser Aspekt soll in künftigen Publikationen dieser Reihe ausführlicher behandelt werden. Das Gebiet befindet sich noch in der Entwicklungsphase, und ein Abschluß ist nicht in Sicht. Wichtige neue Begriffe aus anderen Teilgebieten, wie Quasikristalle, inkommensurable Strukturen und Baufehler, aber auch Fraktale, finden teilweise schon jetzt über verallgemeinerte Symmetrien ihre Einordnung in die umfassende Symmetrielehre der kondensierten Materie.

Auch deshalb sollte der Leserkreis dieses Offset-Titels recht breit sein. Eine höhere Auflage evtl. in englischer Übersetzung wäre sehr zu begrüßen. P. PAUFLER

Cryst. Res. Technol. | 23 | 1988 | 4 | 517–524

I. DÖRFEL

Humboldt-Universität zu Berlin, Sektion Physik

Results of Texture Simulation in b.c.c. Materials

Rolling textures of b.c.c. polycrystals are simulated by the "full constraint" (FC)- and the "relaxed constraint" (RC)-Taylor model, by the FC-model with glide asymmetry in {112} ⟨111⟩ slip systems included, by a combination of FC- and RC-model, and by this combination with glide asymmetry included, respectively. Several cases of activated slip system families have been investigated. Simulation results are compared qualitatively with experimental ones.

Walztexturen von kubisch-raumzentrierten Polykristallen werden mit Hilfe des Taylor-Modells mit vollständigen bzw. abgeschwächten Deformationsbedingungen (FC- bzw. RC-Modell) und durch Kombination und Erweiterung dieser Modelle unter Berücksichtigung der in der Gleitsystemfamilie {112} ⟨111⟩ auftretenden Gleitasymmetrie simuliert. Verschiedene Fälle von aktivierten Gleitsystemfamilien werden untersucht. Es erfolgt ein qualitativer Vergleich der Simulationsergebnisse mit experimentellen Werten.

Introduction

The prediction of deformation textures is an important part in materials science being connected with creating "materials with projected properties". TAYLOR's theory of plastic flow of polycrystals (TAYLOR; VAN HOUTTE 1975) and its further development, the RC-model (VAN HOUTTE 1982) is considered as the basis of calculations. The last model mentioned was proposed, because the classical Taylor model does not take into account the inhomogeneity of real plastic deformation. The textures simulated by that model become sharper than the experimental ones, and their components are shifted. In metallographic experiments often only three or four slip systems activated inside the grain were found. To adjust this the model takes into account the change of grain forms by large deformations and releases the former rigid matching rules.

In this paper textures of the rolling deformation of a niobium polycrystal have been simulated by several models: the classical TAYLOR model, the TAYLOR model with a slip asymmetry in twinning and antitwinning direction, found in the slip systems {112} ⟨111⟩ of b.c.c. materials, the RC-model, a combination of FC- and RC-model, and a combination taking into account the slip asymmetry, too.

For room temperature deformations of niobium polycrystals in theoretical and experimental investigations the slip systems {110} ⟨111⟩ and {112} ⟨111⟩ were found (VAN DER WALT; BOWEN; FRANCIOSI; FRANK, ŠESTÁK). The critical shear stress in both slip system families was assumed to be equal.

Simulations

In the simulations the possible cases of glide on 1.: {110} ⟨111⟩ only, or on 2.: {112} ⟨111⟩ only, or 3.: on both slip system families, respectively, were investigated.

A rolling reduction of 90% during single steps with a displacement gradient for each step of

$$T^{s} = \begin{bmatrix} -\varepsilon & 0 & 0 \\ 0 & 0 & 0 \\ 0 & 0 & \varepsilon \end{bmatrix} \quad (1)$$

with $\varepsilon = 0.05$ was assumed. This tensor means that in the sample system the axis x is the sheet plane normal, axis y is the transverse direction, and axis z is the rolling direction.

The simulation were done for 147 initial crystallites, distributed randomly in the fundamental area of the Euler space (VAN HOUTTE, AERNOUDT; POSPIECH). For cubic crystal symmetry and orthorhombic strain symmetry, the fundamental area is $0° \leqq \Phi \leqq 54.7°$, $0° \leqq \varphi_2 \leqq 45°$ and φ_1 in a range of 180°. On the boundarys of the fundamental area φ_1 is $< 90°$ (VAN HOUTTE, AERNOUDT; DÖRFEL 1985).

A procedure for reflecting orientations, which have left the fundamental area during simulation, was carried out in every simulation step (VAN HOUTTE, AERNOUDT; DÖRFEL 1985). Simulation with the FC-model: The mathematical methods used for this kind of model are well known (VAN HOUTTE 1975). The choice of the activated combination of five slip systems is performed according to TAYLOR's minimum condition by a special simplex algorithm. In the case, when more than one basic solution are found, the simulation uses their average value. Simulation with FC-model, taking into account slip asymmetry: The simulation is similar to that with the pure FC-model. The glide asymmetry in the twinning and antitwinning direction of the {112} ⟨111⟩ slip systems is taken into account in TAYLOR's minimum condition

$$\sum_{k=1}^{5} \tau_{c}^{k} \, |\gamma_{k}| = \text{minimum} \quad (2)$$

with

$\tau_c^k = \tau_c = 1$ for glide systems {110} ⟨111⟩, $\tau_c^k = \alpha \cdot \tau_c$ for glide systems {112} ⟨111⟩ with slip amounts $\gamma_k > 0$ and $\tau_c^k = \beta \cdot \tau_c$ for glide systems {112} ⟨111⟩ with slip amounts $\gamma_k < 0$. For niobium it has been chosen $\alpha = 0.95$ and $\beta = 1.05$ corresponding to experimental results (FOXALL et al.; WASILEWSKI).

$$\left.\begin{aligned} &\left(\alpha = \frac{\tau_c \text{ twinning direct. } \{112\} \langle 111\rangle}{\tau_c\{110\} \langle 111\rangle},\right. \\ &\left.\beta = \frac{\tau_c \text{ antitwin. direct. } \{112\} \langle 111\rangle}{\tau_c \{110\} \langle 111\rangle}\right). \end{aligned}\right\} \quad (3)$$

Simulation with the RC-model: For this simulation the so-called "pancake-model" is used with unconstrained components ε_{21} and ε_{31} of the displacement gradient tensor. As a result (VAN HOUTTE 1982; KOCKS, CHANDRA) three slip systems in a grain will be activated and TAYLOR's minimum condition becomes

$$\sum_{k=1}^{3} \tau_c \, |\gamma_k| = \text{minimum} \, . \quad (4)$$

For calculation of the grain rotation it is necessary to know the magnitude of the free shears ε_{21} and ε_{31}. They can be calculated by solving a system of linear equations with correction terms. T^{g} is the displacement gradient tensor of the grain in system of sample coordinates.

$$T^{g} = \begin{bmatrix} -\varepsilon & 0 & 0 \\ \varepsilon_{21} & 0 & 0 \\ \varepsilon_{31} & 0 & \varepsilon \end{bmatrix} \quad (5)$$

T^g has to be transformed into the system of crystal coordinates and supplies the right hand sides of the system of linear equations. It consists of two parts.

$$T^{gc} = T^{gc}_{(0)} + T^{gc}_{(r)} . \tag{6}$$

The $T^{gc}_{(r)}$ are the correction terms, resulting from the shears ε_{21} and ε_{31}. They have to be symmetrized before being substituted in the matrix of coefficients (DÖRFEL 1986).

$$\sum_{k=1}^{3} \gamma_k \cdot s_{ij^s_k} = T^{gc}_{(0)} + T^{gc}_{(r)} , \tag{7}$$

$$-T^{gc}_{(r_s)} + \sum_{k=1}^{3} \gamma_k \cdot s_{ij^s_k} = T^{gc}_{(0)} . \tag{8}$$

The $T^{gc}_{(r_s)}$ are

$$\left.\begin{aligned}
\varepsilon^{11}_{(21)} &= \varepsilon[(\sin\varphi_1 \cos\varphi_2 + \cos\varphi_1 \sin\varphi_2 \cos\Phi)(\cos\varphi_1 \cos\varphi_2 + \\
&\qquad -\sin\varphi_1 \sin\varphi_2 \cos\Phi)] , \\
\varepsilon^{22}_{(21)} &= \varepsilon[(-\sin\varphi_1 \sin\varphi_2 + \cos\varphi_1 \cos\varphi_2 \cos\Phi) \times \\
&\qquad \times (-\cos\varphi_1 \sin\varphi_2 - \sin\varphi_1 \cos\varphi_2 \cos\Phi)] , \\
\varepsilon^{33}_{(21)} &= \varepsilon[-\cos\varphi_1 \sin^2\Phi \sin\varphi_1] , \\
\left.\begin{matrix}\varepsilon^{12}_{(21)} \\ \varepsilon^{21}_{(21)}\end{matrix}\right\} &= \varepsilon[-\sin^2\varphi_1 \cos^2\varphi_2 \cos\Phi - \cos^2\varphi_1 \sin^2\varphi_2 \cos\Phi + \\
&\qquad -\cos\varphi_1 \sin\varphi_2 \cos^2\Phi \sin\varphi_1 \cos\varphi_2 + \\
&\qquad -\sin\varphi_1 \sin\varphi_2 \cos\varphi_1 \cos\varphi_2 + \tfrac{1}{2}\cos\Phi] , \\
\left.\begin{matrix}\varepsilon^{23}_{(21)} \\ \varepsilon^{32}_{(21)}\end{matrix}\right\} &= \varepsilon[-\sin^2\varphi_1 \sin\varphi_2 \sin\Phi + \cos\varphi_1 \cos\varphi_2 \cos\Phi \sin\varphi_1 \sin\Phi + \\
&\qquad + \tfrac{1}{2}\sin\Phi \sin\varphi_2] , \\
\left.\begin{matrix}\varepsilon^{13}_{(21)} \\ \varepsilon^{31}_{(21)}\end{matrix}\right\} &= \varepsilon[\tfrac{1}{2}\cos\varphi_2 \sin\Phi + \cos\varphi_1 \sin\varphi_2 \cos\Phi \sin\varphi_1 \sin\Phi + \\
&\qquad -\cos^2\varphi_1 \sin\Phi \cos\varphi_2]
\end{aligned}\right\} \tag{9}$$

$$\left.\begin{aligned}
\varepsilon^{11}_{(31)} &= \varepsilon[\sin\varphi_2 \sin\Phi(\cos\varphi_1 \cos\varphi_2 - \sin\varphi_1 \sin\varphi_2 \cos\Phi)] , \\
\varepsilon^{22}_{(31)} &= \varepsilon[\cos\varphi_2 \sin\Phi(-\cos\varphi_1 \sin\varphi_2 - \sin\varphi_1 \cos\varphi_2 \cos\Phi)] , \\
\varepsilon^{33}_{(31)} &= \varepsilon[\cos\Phi \sin\varphi_1 \sin\Phi] , \\
\left.\begin{matrix}\varepsilon^{12}_{(31)} \\ \varepsilon^{21}_{(31)}\end{matrix}\right\} &= \varepsilon[-\sin^2\varphi_2 \sin\Phi \cos\varphi_1 - \sin\varphi_2 \sin\Phi \sin\varphi_1 \cos\varphi_2 \cos\Phi + \\
&\qquad + \tfrac{1}{2}\sin\Phi \cos\varphi_1] , \\
\left.\begin{matrix}\varepsilon^{23}_{(31)} \\ \varepsilon^{32}_{(31)}\end{matrix}\right\} &= \varepsilon[-\tfrac{1}{2}\cos\Phi \cos\varphi_1 \sin\varphi_2 - \sin\varphi_1 \cos\varphi_2 \cos^2\Phi + \\
&\qquad + \tfrac{1}{2}\cos\varphi_2 \sin\varphi_1] \\
\left.\begin{matrix}\varepsilon^{13}_{(31)} \\ \varepsilon^{31}_{(31)}\end{matrix}\right\} &= \varepsilon[-\cos^2\Phi \sin\varphi_1 \sin\varphi_2 + \tfrac{1}{2}\cos\Phi \cos\varphi_1 \cos\varphi_2 + \\
&\qquad + \tfrac{1}{2}\sin\varphi_1 \sin\varphi_2] .
\end{aligned}\right\} \tag{10}$$

The choice of the activated slip system combination is performed as in the FC-model by a special simplex algorithm, but when more than one solution fulfills the minimum condition instead of an average value the solution, with its sum of the amounts of the free shears being a minimum, is selected.

$$H = |x_{21}| + |x_{31}| = \text{minimum} . \tag{11}$$

Simulation with a combination of models: in connection with forming non-equiaxed grains by heavily cold-rolled polycrystals, a combination of the FC- and RC-models is used for their simulation. For niobium with a rolling degree of 20% (THOMPSON, FLEWITT), flat, disc-shaped grains were found in the whole material. To simulate this behaviour, the FC-model is used up to about 20% rolling degree, and these results are the start orientations for additional simulations with the RC-model. Simulation with combination of models and consideration of glide asymmetry: The simulation is performed with the prescribed combination of the FC- and RC-models, in both parts glide asymmetry in the slip system family {112} ⟨111⟩ is considered.

Results and discussion

To compare the simulation results we use experiments from the CINR Rossendorf obtained via neutron diffraction at a sample of polycrystalline niobium (technically pure, ≈ 98% niobium). The sample has been deformed at room temperature using a rolling degree of 90%. This texture is shown in Figure 1. It consists of a line of stable end orientations with directions of the crystal axes from ⟨011⟩ to ⟨123⟩ and ⟨112⟩ in rolling direction with the 2- to 3-fold of the random distribution in ⟨123⟩ and ⟨112⟩ and with the 8-fold of the random distribution in ⟨011⟩. In the rolling plane there are crystal planes {001} and {111} with the 8- and 6-fold of the random distribution, respectively. All simulations have been performed up to 90% rolling degree and the arising textures are figured in inverse pole figures of the rolling direction and its components in percentages, shown in tables. The calculations have been performed in the Humboldt-University computer centre (DÖRFEL 1987).

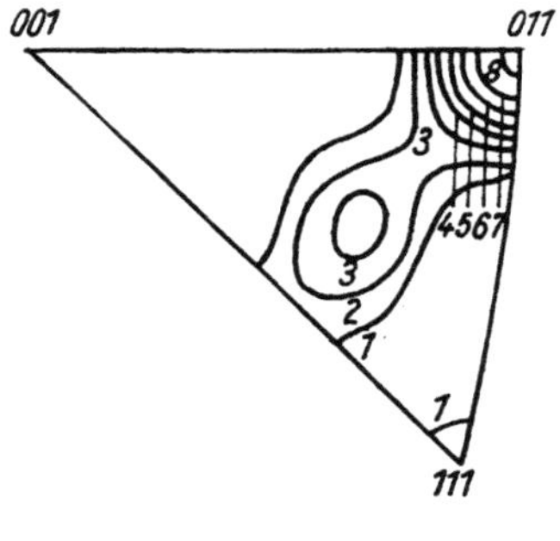

Fig. 1. Experimental rolling texture of niobium, 90% rolling degree inverse pole figure of rolling direction.

In the FC-model-simulation at 60% rolling degree, stable end orientations are already distinct. The results for 90% in cases 1, 2 and 3 (only {110} ⟨111⟩, only {112} ⟨111⟩, both slip system families) are shown in Tables 1a, 1b, 1c and in Figures 2a, 2b, 2c. In case 1 there are three areas of stable end orientations: with directions of crystal axes between ⟨113⟩ and ⟨124⟩, ⟨011⟩ and ⟨0vw⟩ in the rolling direction.

In case 2 there is an area with large intensity around ⟨011⟩ {100} and weaker areas around ⟨136⟩ {$\bar{1}\bar{3}2$} and with ⟨0vw⟩ in rolling direction.

In case 3 there are three areas of stable end orientations: near ⟨011⟩ in rolling direction, an area with ⟨0vw⟩ in rolling direction and a line from ⟨225⟩ to ⟨124⟩ and ⟨157⟩ in rolling direction.

The component ⟨011⟩ in rolling direction is in case 2 most intensive, but in this case the orientation line is degenerated to a component ⟨136⟩ {$\bar{1}\bar{3}2$}. In case 1 and 3 the orientation line is shifted from ⟨112⟩ in the experiment to orientations near by ⟨113⟩ in simulations. The components with ⟨011⟩ in rolling direction are to weak,

Table 1a
Simulation of rolling texture by FC-model, case 1 (only {110} ⟨111⟩)

Texture components	average Euler angles/°			percentage/%
	Φ	φ_2	φ_1	
⟨011⟩ {1$\bar{2}$1}, {111}, {75$\bar{5}$}	42.7	0.2		22.0
⟨113⟩ {$\bar{3}\bar{3}$2}, {5$\bar{7}\bar{9}$}, {012}	27.9	35.5		29.3
⟨124⟩ {$\bar{1}$12}	30.5	22.0	−58.0	6.1
⟨124⟩ {86$\bar{5}$}	28.8	28.5	−67.0	12.9
⟨014⟩ {04$\bar{1}$}, {01$\bar{3}$}	15.8	0.0		4.7
⟨012⟩ {02$\bar{1}$}	25.9	0.0	−89.9	6.8
⟨023⟩ {1 12 $\bar{9}$}	35.3	0.2	−86.4	5.4

Table 1b
Simulation of rolling texture by FC-model, case 2 (only {112} ⟨111⟩)

Texture components	average Euler angles/°			percentage/%
	Φ	φ_2	φ_1	
⟨011⟩ {557}, {112}, {113}, {100}, {0$\bar{1}\bar{1}$}	43.8	0.4		73.0
⟨136⟩ {$\bar{1}\bar{3}$2}	28.8	16.3	90.5	10.0
⟨012⟩ {02$\bar{1}$}	26.2	0.2	−85.6	10.0
⟨015⟩ {011}, {100}	11.7	0.0		

Table 1c
Simulation of rolling texture by FC-model, case 3 (both slip system families)

Texture components	average Euler angles/°			percentage/%
	Φ	φ_2	φ_1	
⟨011⟩ {0$\bar{1}$1}, {100}, {31$\bar{1}$}, {21$\bar{1}$}, {32$\bar{2}$}	43.7	0.2		35.0
⟨124⟩ {21$\bar{1}$}, {75$\bar{4}$}	29.4	27.9	−66.6	16.3
⟨57 16⟩ {75$\bar{4}$}, {33$\bar{2}$}	27.8	36.4	−78.4	12.9
⟨157⟩ {21$\bar{1}$}, {210}	40.0	12.0	−44.5	10.9
⟨113⟩ {33$\bar{2}$}	27.4	42.4	−86.2	6.8
⟨012⟩ {02$\bar{1}$}	24.4	0.0	−89.8	4.8

Table 2
Simulation of rolling texture by special FC-model with glide asymmetry included, case 3

Texture components	average Euler angles/°			percentage/%
	Φ	φ_2	φ_1	
⟨011⟩ {0$\bar{1}$0}, {21$\bar{1}$}, {41$\bar{1}$}	44.1	0.5		22.5
⟨158⟩ {21$\bar{1}$}	34.2	13.7	−48.7	13.6
⟨034⟩ {21$\bar{1}$}	37.9	6.9	−39.5	4.0
⟨124⟩ {21$\bar{1}$}	30.9	21.2	−57.5	6.8
⟨238⟩ {65$\bar{4}$}	27.9	34.8	−75.8	18.4
⟨113⟩ {33$\bar{2}$}	27.5	40.7	−84.1	18.4
⟨124⟩ {75$\bar{4}$}	29.2	27.0	−65.9	10.9

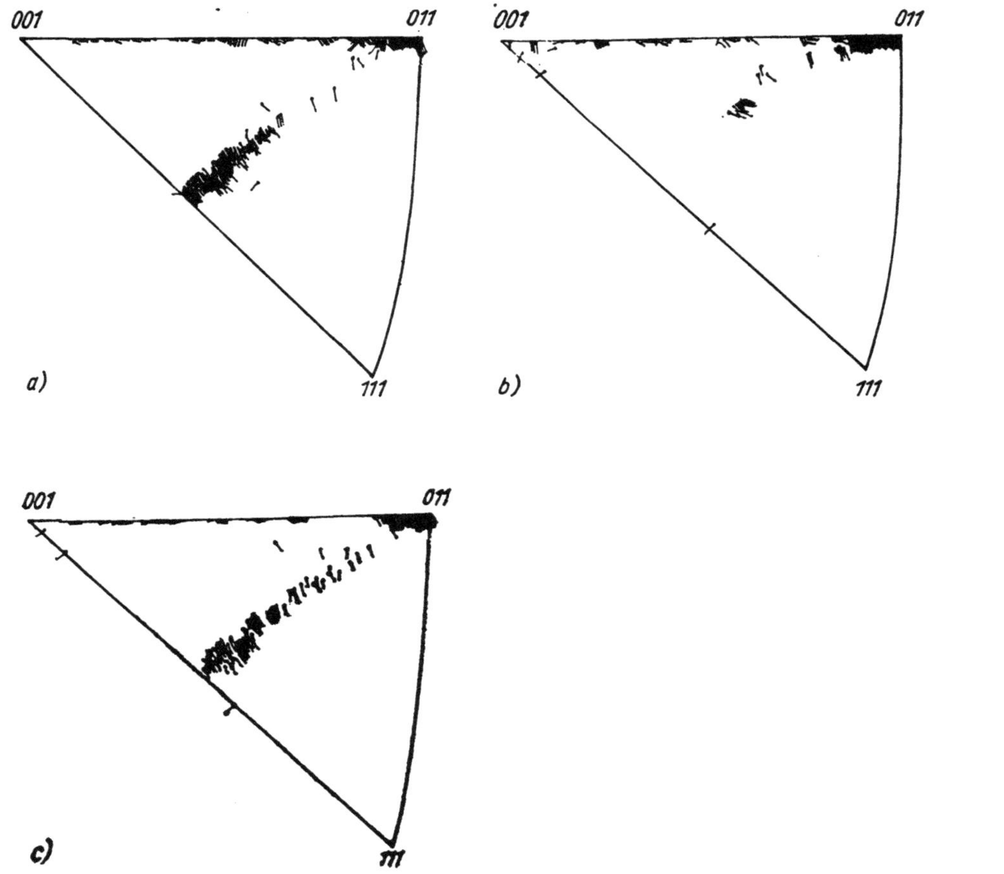

Fig. 2a. Rolling texture from FC-model-simulation, case 1 inverse pole figure of rolling direction, dashes give the normal direction; 2b. Rolling texture of FC-model-simulation case 2, inverse pole figure of rolling direction, dashes give the normal direction; 2c. Rolling texture by FC-model-simulation, case 3, inverse pole figure of rolling direction, dashes give the normal direction

and in all cases components ⟨0vw⟩ exist in rolling direction, which have not been found in experiments. In the experiments the planes {100} and {111} are with large intensity in the rolling plane, in simulations {100}-component is only weak and no planes occure exactly parallel to {111} in rolling plane, but such with about 10° distance from {111}.

In simulation with the special FC-model, where the glide asymmetry has been considered, the formation of stable end orientations begins at the same rolling degree like in pure FC-model. The influence of glide asymmetry creates other end orientations, shown in Table 2 and Figure 3. Only the interesting case 3 is presented. Two areas of stable end orientations have been created: one with crystal directions ⟨011⟩ in rolling direction and another between ⟨113⟩ to near ⟨011⟩ in rolling direction. The component ⟨011⟩ in rolling direction is in simulation weaker than in experiments, but the components ⟨0vw⟩, disturbing in simulation by pure FC-model, are absent here.

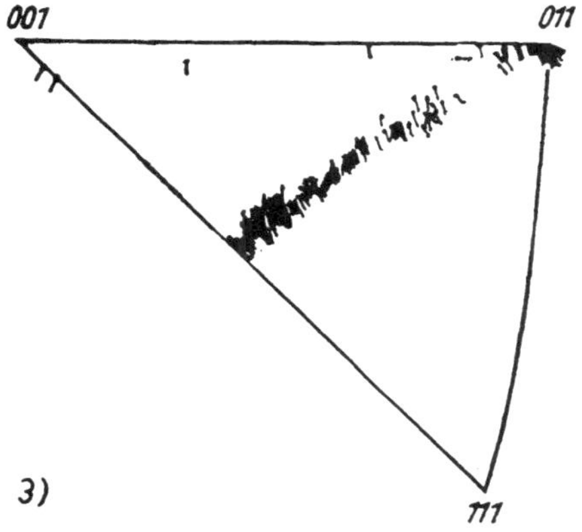

3)

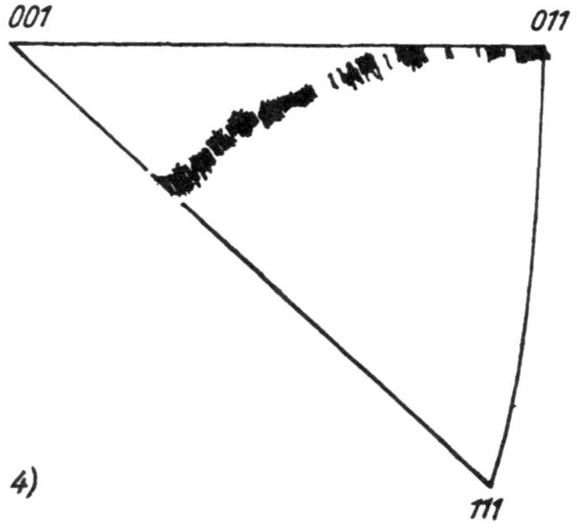

4)

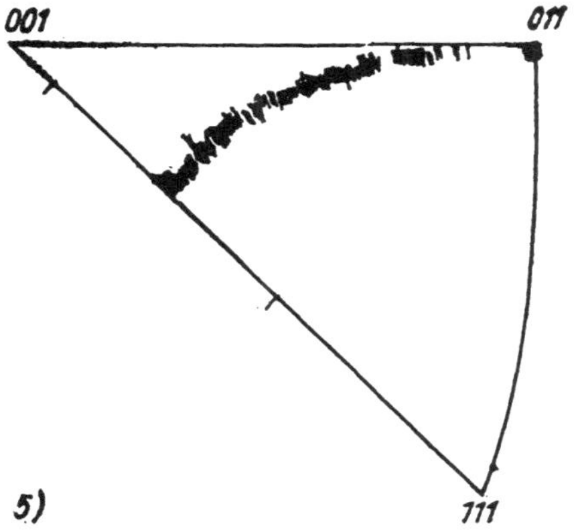

5)

Fig. 3. Rolling texture simulation by special FC-model with glide asymmetry in {112}⟨111⟩, case 3, inverse pole figure of rolling direction, dashes give the normal direction

Fig. 4. Rolling texture simulation by RC-model, case 1, inverse pole figure of rolling direction, dashes give the normal direction

Fig. 5. Rolling texture simulation by a combination of FC- and RC-model with glide asymmetry included, case 3, inverse pole figure of rolling direction, dashes give the normal direction

The results of simulations on the basis of the RC-model, a combination of FC- and RC-model and with additional glide asymmetry at {112} ⟨111⟩-slip systems included are worse. In all cases the formation of stable end orientations can be seen already at 30% rolling degree. At 90% rolling degree the texture is very sharp and in some cases the orientation line seems to decay into single orientations. This can be seen best in the results of RC-model, case 1 (Fig. 4). Orientation lines, stretching in experiments from ⟨112⟩ to ⟨011⟩ in rolling direction, are significantly shifted to ⟨114⟩ and in some cases they reach components near by ⟨023⟩ in rolling direction instead of ⟨011⟩ (combination with asymmetry, case 3), (Fig. 5). In those cases components ⟨011⟩ are clearly separated from the orientation line and very weak. The components with ⟨011⟩-direction in rolling direction in most simulations with RC-model and its combination are of small intensity, percentage of that components are about 10%. Only in simulations by RC-model and by a model combination, in cases 2 percentages for ⟨011⟩-components reaches 30%. The distribution of crystal planes in the rolling plane is satisfactory in simulation results of RC-model cases 2 and 3, of model combination case 3 and of model combination with glide asymmetry included. Planes {100} and {111} are in the rolling plane. Planes exactly parallel to {111} have not been found.

Conclusions

In this work rolling textures of a b.c.c. polycrystal (special behaviour of niobium) have been considered. Being compared qualitatively with experimental data, the results of the simulation by means of the FC-model, with a glide asymmetry in {112} ⟨111⟩-slip systems included, show the best agreement. Indeed the components with ⟨011⟩ in rolling direction are weaker as in the experiments, but the orientation line is shifted only weakly, no disturbing components with ⟨0vw⟩ in rolling direction are present and planes in rolling plane are not so far from the experimental ones. The agreement is not yet satisfactory.

Simulation results by RC-model or combination from FC- and RC-model are still worse. There the texture is very sharp, components are significantly shifted and components with ⟨011⟩ in rolling direction are to weak.

For better agreements the FC-model should be recalculated with a glide asymmetry and with weighting factors for the cases 1 and 2. The glide asymmetry constant should be choosen at the same level as before.

The author is indebted to M. Schenk, P. Bankwitz, and M. Betzl for scientific and material support and G. Schnabel for programming.

References

Bowen, D. K. et al.: Can. J. Phys. **45** (1967) 903
Dörfel, I.: Communication of the JINR Dubna 1985, P17-85-886 (in Russ.)
Dörfel, I.: Communication of the JINR Dubna 1986, P17-86-683 (in Russ.)
Dörfel, I.: Ph. D. Thesis, Humboldt-University Berlin 1987
Foxall, R. A. et al.: Can. J. Phys. **45** (1967) 607
Franciosi, P.: Acta Metallurg. **31** (1983) 1331
Frank, W., Šesták, B.: Skripta Met. **4** (1970) 451
Kocks, U. F., Chandra, H.: Acta Metallurg. **30** (1982) 695
Pospiech, J.: Kristall u. Technik **7** (1972) 1057
Taylor, G. I.: J. Inst. Metals **62** (1938) 307
Thompson, S. J., Flewitt, P. E.: J. Less-Common Metals **40** (1975) 269
van der Walt, C. M.: Acta Metallurg. **17** (1969) 393
van Houtte, P.: Z. Metallkde. **66** (1975) 202
van Houtte, P., Aernoudt, E.: Mat. Sci. Engin. **23** (1976) 11
van Houtte, P.: Mat. Sci. Engin. **55** (1982) 69
Wasilewski, R. J.: Scripta Metall **1** (1967) 45

(Received August 6, 1987;
accepted September 24, 1987)

Author's address:

I. Dörfel
Humboldt-Universität zu Berlin, Sektion Physik, Bereich Kristallographie
Invalidenstr. 110
Berlin
DDR-1040

Cryst. Res. Technol. | 23 | 1988 | 4 | 525–529

Y. G. Zargaryan, P. H. Bezirganyan

Yerevan State University, USSR

Peculiarities of Pendellösung Fringes in Quartz Monocrystals

The scattering of X-rays in quartz monocrystals has been investigated. It has been discovered experimentally, that even for the thick quartz crystals, for which $\mu t = 5$, pendular bands can be obtained. The appearance of pendular bands depending on the crystal thickness, the reflecting plane families and the reflection order, was investigated as well.

Исследовано рассеяние рентгеновских лучей в кварцевых монокристаллах. Экспериментально обнаружено, что даже в таких толстых кристаллах кварца, для которых $\mu t = 5$ возникают маятниковые полосы. Исследовано также возникновение маятниковых полос в зависимости от семейства отражающих плоскостей и порядка отражения.

As it is known (Ewald; Laue) the pendular distribution of X-ray intensity in crystals is due to interference taking place between the waves $\boldsymbol{K}_{01}$ and $\boldsymbol{K}_{02}$ diffracted along the incidence direction and between the waves $\boldsymbol{K}_{h1}$ and $\boldsymbol{K}_{h2}$ diffracted along the reflection direction (in two-field approximation, Fig. 1). On the other side it is known also that the first field (waves $\boldsymbol{K}_{01}$ and $\boldsymbol{K}_{h1}$) in sufficiently thick crystals ($\mu t > 1$ where t is the thickness of the crystal, μ the absorption coefficient) is almost completely absorbed due to anomalous absorption, and, thus, the pendular energy distribution of refracted

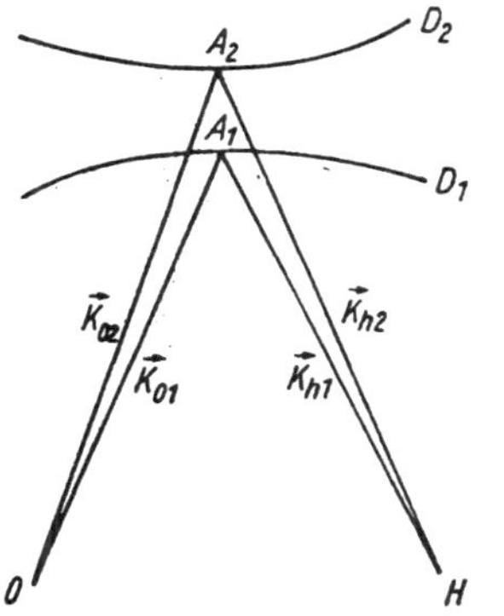

Fig. 1. Representation scheme of the wave-field in crystals

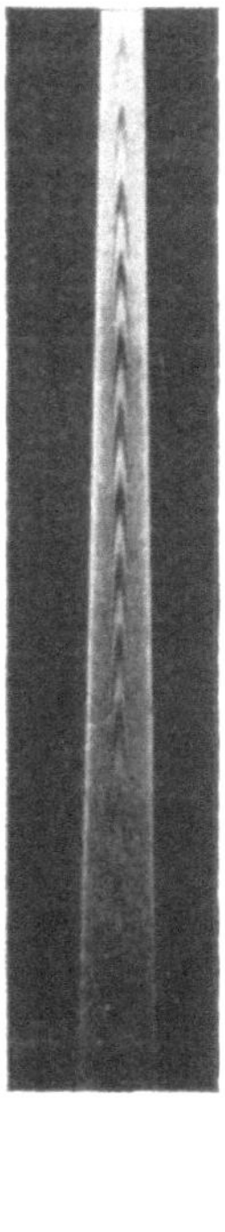

Fig. 2. Pendular bands from thick quartz crystals ($\mu t \geqq 10$)

X-rays can occur only in thin crystals for which $\mu t \leqq 1$. Meanwhile, our experimental investigations have shown a clear picture of pendular bands obtained from a thick wedge-like quartz crystal, which comes to prove that in thick quartz crystals pendular intensity distribution of refracted waves occur ($\mu t = 4.5$ in the middle part of the crystal, meanwhile for Si crystals $\mu t \leqq 1.5$).

For interpretation of this unexpected effect it is reasonable to suppose that in quartz crystals under investigation the anomalous absorption (the Bormann effect) is absent, and both fields are absorbed identically. Thus, in thick ($\mu t \geqq 4.5$) quartz crystals pendular distribution occurs, and we observed pendular bands out of the crystal. Thus, one may conclude, that in thick quartz crystals the anomalous absorption (the Bormann effect) may be absent due to imperfect structure, but pendular bands can occur. In order to be convinced in this we had to carry out additional experimental investigations.

In our experiments we assumed that the Bormann effect can occur only when the energy of the first field in the crystal flows along the reflecting planes (anomalous large absorption), and the energy of the second field propagates between these planes (anomalous small absorption) which is due to interference between waves of the first field ($\boldsymbol{K}_{01}$ and $\boldsymbol{K}_{h1}$) and of the second field ($\boldsymbol{K}_{02}$ and $\boldsymbol{K}_{h2}$).

To prove this assumption we developed a method, which gave the possibility to observe visually the direction of refracted X-ray wave energy flows in crystals depending on size and perfectness of these crystals (BEZIRGANYAN et al. 1984a, b). As during the kinematic scattering the energy of X-ray waves in crystal flows along the incidence and reflection directions (Fig. 3a), and during the dynamic scattering it flows along the reflecting planes, so the cross-sections of X-ray wave beams, refracted in step-like crystals (see BEZIRGANYAN et al. 1984) will have the form shown in Figures 4a and 4b, respectively; during kinematic scattering the beam cross section has the form of reflecting planes (Fig. 4a) and from one side it is smooth, but from the other one it's stepped, whereas at the dynamic scattering (Fig. 4b) parts of refracted beam tracks are shifted with respect to one another, the thickness of these tracks being identical (in fact almost identical). In all these cases the preliminary beam fell from the smooth side of the crystal (BEZIRGANYAN et al. 1984).

Using the mentioned above method, the energy flows and the production of pendular bands of refracted waves in quartz crystals was investigated.

It is shown that

1. even in sufficiently thick quartz crystals the energy of waves flows along the incidence and reflection directions (Fig. 5), i.e. in these crystals the Bormann effect is absent.

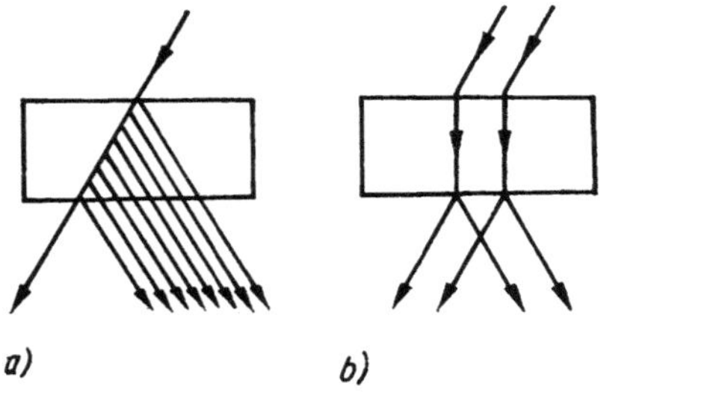

Fig. 3. The rays-traces in the crystal during reflection: a) kinematic scattering; b) dynamic scattering

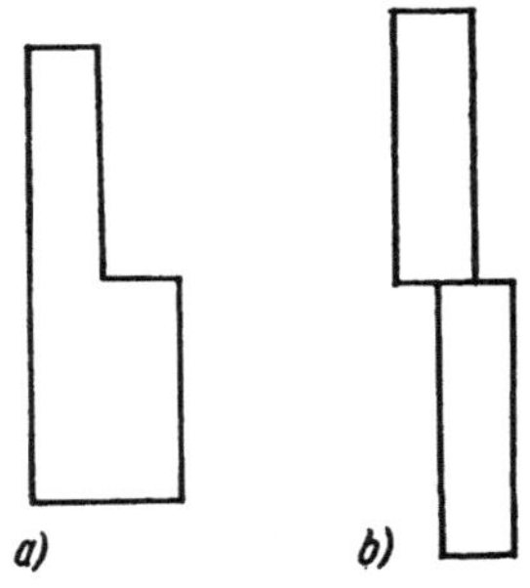

Fig. 4. The cross-section of reflected beam: a) kinematic reflection; b) dynamic reflection

2. therefore, even in thick quartz crystals ($\mu t = 5$) pendular distribution of energy occurs, and in such thick wedge-like crystals a sufficiently clear picture of pendular bands is observed (see Fig. 2).
3. for more convincing proof of the fact that in one and the same sufficiently thick crystal the kinematic scattering takes place (energy flow along the incidence and reflection directions), and, simultaneous by, pendular distribution of the energy of refracted waves occurs, a step-wedge-like crystal was made (see Fig. 6a), off which a picture of pendular bands is obtained (Fig. 6b). As it is seen in this picture the cross section of diffracted beam from the left side is smooth, and in both upper and lower parts of the figure the pendular bands are seen.

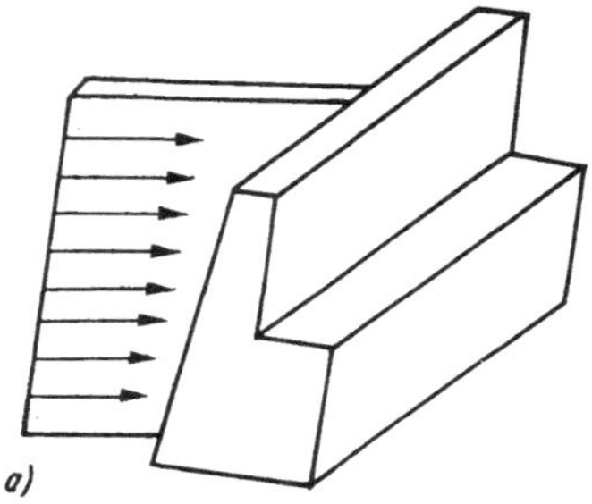

a)

Fig. 6. a) Step-wedge-like crystal. Preliminary beam falls from smooth side of the crystal; b) Pendular bands obtained from step-wedge-like crystal

◀ Fig. 5. Cross-section of X-ray beam, refracted in wedge-shaped quartz crystals

b)

Thus, we consider, proved, that

a) in sufficiently thick quartz crystals kinematic scattering takes place.
b) and most important, in thick quartz crystals ($\mu t \geqq 10$) where the Bormann effect is absent (i.e. energy flow along the incidence and reflection directions), pendular distribution of diffracted wave energy occur.

Further, the appearance of pendular bands depending on the crystal thickness, reflecting planes families and reflection order was investigated.

It is shown that the stronger reflection, the smaller the maximum thickness of the crystal, from which pendular bands can be still received. It is also shown that this maximum thickness increases with the increase of the reflection order. For example, from strongly reflecting planes (10$\bar{1}$1) the pendular bands are obtained when $\mu t \leqq 4$, but at the reflection from weakly reflecting planes (10$\bar{1}$0), (11$\bar{2}$0), (01$\bar{1}$0), the pendular bands are obtained when $\mu t \leqq 5.3$. The last three families bring to pendular bands in the second reflection order if $\mu t \leqq 6.1$. From our point of view the increase of maximum thickness (still yielding pendular bands) with the decrease of reflecting intensity, is due to the anomalous absorption with the decrease of structure factor, otherwise, with weakening of the dynamic interaction between the refracted waves.

Then an interesting investigation of the dependence of direction of V-shaped pendular bands on the order of refraction of X-rays is carried out. It is known (HART,

MILNE) that in two-crystal HART-MILNE system the tops of the pendular bands are directed to the wide part of the wedge, and by BEZIRGANYAN, and BEZIRGANYAN et al. 1976 it was shown that when the parallele piped-shaped crystal is thin then the tops of the pendular bands are directed to the narrow part of the wedge. Our investigations have shown that except for this the direction of tops of the pendular bands in quartz crystals depends also on the order of refraction of X-rays. Indeed the two-crystal system was made from quartz crystals, which consisted of parallelepiped and wedge-shaped crystals. The reflecting planes were the R(10$\bar{1}$1) planes of "x" cut. The highest thickness of the parallelepiped crystal was adjusted to have tops of the pendular bands of first order of reflection to be directed to the wide part of the wedge (see Fig. 7). In Figure 7a the pendular bands behind the wedge are shown. One can see in this figure that the tops of these bands are directed to the narrow part of the wedge

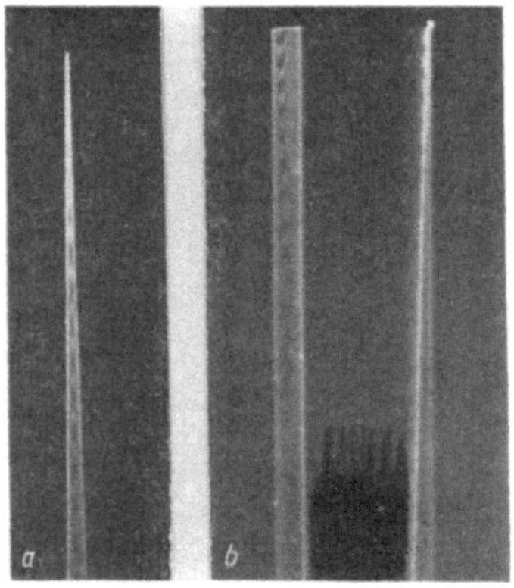

Fig. 7. The X-ray topograph from the two-crystal system of the first order of reflection: a) behind the wedge; b) behind the parallelepiped crystal (behind the system)

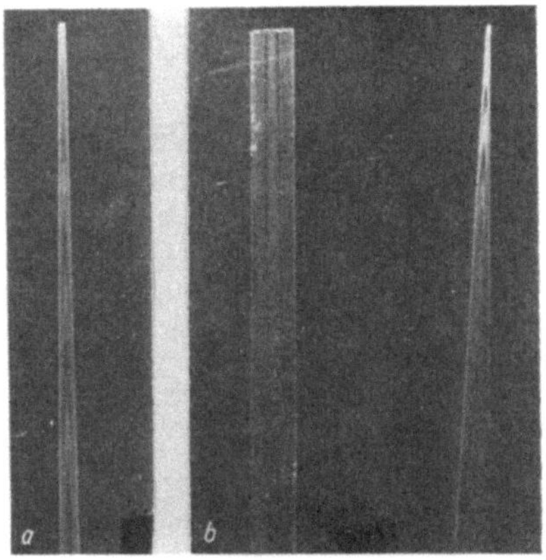

Fig. 8. The X-ray topograph from the two-crystal system of the second order of reflection: a) behind the wedge; b) behind the parallelepiped crystal (behind the system)

(an incident wave goes from the wedge direction). In Figure 7b the pendular bands behind parallelepiped crystal are shown (behind the system). One can see in this figure that the pendular bands are reversed, i.e. the tops are directed to the wide part of the wedge, i.e. in the first order of reflection in flat parallelepiped block tops of the pendular bands are reversed in other direction. Then using the same two-crystal system the pendular bands of the second order are obtained (Fig. 8). In Figure 8a the pendular bands obtained behind the wedge are shown. One can see in this figure that in the second as well as in the first orders of reflection the V-shaped pendular bands are obtained behind the wedge, the tops of these bands are directed to the narrow part of the wedge. But in this case the angle of divergence is much larger. In Figure 8b the pendular bands behind the parallelepiped block are shown. One can see in this figure that these bands are not reversed: tops are directed to the narrow part of the wedge.

Thus we come to an important conclusion: in a system consisting of one wedge- and one parallelepiped-shaped crystal a thickness of the last crystal can be adjusted so as the V-shaped pendular bands formed in the wedge-shaped crystal, after reflecting in the parallelepiped crystal in the first order of reflection were directed to the wide part of the wedge and in the second order of the reflection, that is, to the narrow part of the wedge. This result can be explained on the following speculative base. By BEZIRGANYAN and BEZIRGANYAN et al. 1976) it was shown that the larger was the angle between the wings of V-shaped band, the deeper distance it took to reverse this band. For reversing of the pendular bands of the second order of reflection the thickness of our parallelepiped crystal was not sufficient, therefore in the first order of reflection the bands are reversed (the thickness in sufficient for that) but in the second order of reflection the bands are not reversed. With an increase of the thickness of the parallelepiped crystal the pendular bands can be reversed also in the second order of reflection.

References

BEZIRGANYAN, P. H.: pre-print. Yerevan State University, Solid State Physics 4—7, 1976
BEZIRGANYAN, P. H., ALADZHADZHYAN, G. M., SEMERDZHYAN, H. S.: Deponent VINITI, No. 4000, 1976
BEZIRGANYAN, P. H., ZARGARYAN, Y. G., ASLANYAN, V. G.: Izv. Akad. Nauk Armen. SSR, Physika **19** (1984)a 35
BEZIRGANYAN, P. H., ZARGARYAN, Y. G., ASLANYAN, V. G.: Cryst. Res. Technol. **19** (5) (1984)b 517
EWALD, P. P.: Ann. Phys. **49** (1916) 117
HART, M., MILNE, H. D.: phys. stat. solidi **26** (1) (1968) 185
LAUE, M.: Röntgenstrahlinterferenzen, Frankfurt am Main 1960

(Received August 17, 1987;
accepted September 10, 1987)

Authots' address:

Y. G. ZARGARYAN, Prof. P. H. BEZIRGANYAN
Yerevan State University, Chair of Solid State Physics
Mravyan, 1
375049 Yerevan-49
Armenia, USSR

Book Review

H. MILITZER, F. WEBER (Hrsg.)

Angewandte Geophysik

mit Beiträgen von B. FORKMANN, H. MILITZER, R. RÖSLER, R. SCHMÖLLER und F. WEBER

Gemeinschaftspublikation im Springer-Verlag Wien, New York (ISBN 3-211-81799-9) und Akademie-Verlag Berlin (ISBN 3-05-500271-7), 1987, 420 Seiten mit 239 Abbildungen und 9 Tabellen, Geb. ÖS 830,–, DM 119,–

Das vorliegende Lehrbuch stellt den 3. und letzten Band des Werkes zur „Angewandten Geophysik" dar. Es schließt in Inhalt und Form eine seit Jahrzehnten bestehende Lücke auf dem Hauptgebiet der angewandten Geophysik, der Seismik. Der Inhalt umspannt Theorie, Methodik, Technik und Anwendung der angewandten Seismik, wobei die mathematisch-theoretischen Grundlagen nur in unbedingt erforderlichem Umfange behandelt werden. Umso umfangreicher werden — völlig berechtigt — Methodik und Anwendung dargelegt. Der Schwerpunkt liegt bei der Suche und Erkundung von Erdöl- und Erdgaslagerstätten. Sehr gut verständlich werden in diesem Zusammenhang die jüngsten Erkenntnisse der Informationstheorie und elektronischen Rechentechnik bei der seismischen Datenverarbeitung vermittelt. Auch bei der umfassenden und gründlichen Behandlung der Interpretationsprozesse wird stets die Zielstellung der Autoren sichtbar, die methodischen Aspekte jeweils in Verbindung zu praxisspezifischen Problemstellungen abzuleiten. Dieser lobenswerte Aspekt wird besonders deutlich in den abschließenden Kapiteln „Geologisch-lagerstättenkundliche Interpretation seismischer Ergebnisse und Beispiele" (F. WEBER, Montanuniversität Leoben) und „Ausgewählte Spezialverfahren der Seismik" (H. MILITZER, Bergakademie Freiberg). Die Aktualität der Ausführungen zeigt sich in diesen Beiträgen u. a. darin, daß nicht nur die Anwendung seismischer Verfahren bei der Lösung struktureller Aufgaben, sondern auch der Suche stratigraphischer Fallen vorgelegt wird (Seismostratigraphie). In knapper Form werden Messungen im Steinkohlen- und Braunkohlenbergbau sowie in der Baugrunduntersuchung skizziert.

Eine umfangreiche Literaturübersicht und ein detailliertes Sachverzeichnis runden das Lehr- und Fachbuch ab.

Das Buch kann uneingeschränkt sowohl als Lehrstoff für Studierende der Geophysik als auch für Fachwissenschaftler auf dem Gebiet der geologischen Suche und Erkundung empfohlen werden.

G. OLSZAK

Cryst. Res. Technol.	23	1988	4	531–536

H. NEUMANN

Sektion Physik, Karl-Marx-Universität Leipzig

Bulk Modulus — Volume Relationship in Alkali Halides with Rocksalt Structure

It is shown that the bulk modulus of the alkali halides with rocksalt structure follows the simple relation $B = b_0(1 - b_1 f_c)\, V_0^{-n}$ where V_0 is the unit cell volume, f_c the spectroscopically defined covalency of the chemical bond, and b_0, b_1, and n are constants having the same values for all compounds. This result is compared with the bulk modulus scaling law found for the tetrahedral semiconductors.

Es wird gezeigt, daß der Kompressionsmodul der Alkalihalogenide mit Kochsalzstruktur durch die einfache Beziehung $B = b_0(1 - b_1 f_c)\, V_0^{-n}$ beschrieben werden kann, wobei V_0 das Volumen der Einheitszelle, f_c die spektroskopisch definierte Kovalenz der chemischen Bindung und b_0, b_1 and n Konstanten sind, die für alle Verbindungen die gleichen Werte haben. Dieses Ergebnis wird mit dem Skalierungsgesetz für den Kompressionsmodul in den tetraedrischen Halbleitern verglichen.

1. Introduction

In recent years, various attempts have been made to generalize the trends in the bulk modulus of the alkali halides with rocksalt structure. In considerations based on the sum rule of BROUT and a relationship derived by SZIGETI the close relation between the bulk modulus and the frequencies of the lattice vibrational modes has been established (MITRA, MARSHALL). However, these models gave only moderate agreement with experiment. A better description of the bulk modulus trends seems to be possible by means of the simple scaling law

$$B = B_0 V_0^{-n}\,, \tag{1}$$

where V_0 is the unit cell volume and B_0 and n are constants. However, even the values for the exponent n in (1) differ from author to author. Assuming that B_0 and n have the same values for all alkali halides, MITRA and MARSHALL determined $n = 1$ including bulk modulus data for all compounds, whereas SIRDESHMUKH and SUBHADRA found $n = 1.06$ considering the change in the bulk modulus with volume in the sodium and rubidium halides. On the other hand, it has been argued by JAYARAMAN et al. that $n = 1.13$ but that B_0 is a constant only for compounds with the same cation. Recently, in analysing the trends in the bulk modulus of the tetrahedrally coordinated elemental and binary semiconductors it has been established that the bulk modulus of all these compounds can be well described by means of relation (1) with an empirically determined value of $n = 1.204$ for the exponent and with a factor B_0 decreasing with the spectrocopically defined bond ionicity f_i according to

$$B_0 = b_0(1 - b_1 f_i)\,, \tag{2}$$

where b_0 and b_1 are constants having the same values for the whole compound family (NEUMANN). Qualitatively, this result agrees with predictions which can be made on the basis of general considerations (COHEN; MAJEWSKI, VOGL). In the present work it is shown that the bulk modulus trends in the alkali halides can also be well described if an ionicity dependence of the factor B_0 in relation (1) is allowed for.

2. Bond ionicity of the alkali halides

According to previous estimates the bond ionicities of the alkali halides with rocksalt structure vary in the rather narrow range $f_i = 0.889 - 0.960$ (LEVINE; SRIVASTAVA; VAN VECHTEN). Thus it is obvious that rather accurate values for the bond ionicities are required in order to evaluate the ionicity dependence of the bulk modulus. For this reason, we have re-evaluated the bond ionicities of all alkali halides following the same procedure as described by LEVINE, but using more recent data for the long-wavelength electronic dielectric constants ε_0 reported by WEMPLE. For the lattice parameters a we use the values adopted by LANDOLT-BÖRNSTEIN as the most reliable ones. The input parameters ε_0 and a of the calculations as well as the resulting values for the homopolar and heteropolar gap energies E_h and C, respectively, and the ionicities f_i are compiled in Table 1. The results of LEVINE are given for comparison. We see

Table 1
Bond ionicity of the alkali halides with rocksalt structure

Compound	ε_0	a/nm	E_h/eV	C/eV	f_i	f_i (LEVINE)
LiF	1.92	0.40262	7.01	22.60	0.912	0.914
LiCl	2.70	0.51399	3.83	11.72	0.904	0.903
LiBr	3.00	0.5501	3.23	9.94	0.904	0.896
LiI	3.40	0.6012	2.59	7.98	0.904	0.890
NaF	1.74	0.46234	4.97	20.34	0.944	0.946
NaCl	2.33	0.56401	3.04	11.50	0.935	0.936
NaBr	2.60	0.59730	2.64	9.81	0.933	0.933
NaI	2.98	0.64728	2.16	7.86	0.930	0.929
KF	1.84	0.5344	3.47	15.61	0.953	0.954
KCl	2.17	0.62929	2.32	10.48	0.953	0.951
KBr	2.35	0.65982	2.06	9.31	0.953	0.953
KI	2.63	0.70656	1.74	7.74	0.952	0.948
RbF	1.86	0.56516	3.02	14.18	0.956	0.955
RbCl	2.17	0.65898	2.07	9.77	0.957	0.956
RbBr	2.34	0.6889	1.85	8.79	0.958	0.955
RbI	2.59	0.7342	1.58	7.43	0.957	0.954

that in most cases the deviations between our f_i values and those of LEVINE are relatively small except for LiBr and LiI where our values are distinctly larger.

From Table 1 it follows that compounds with the same cation have practically the same bond ionicity, the only exceptions being LiF and NaF. Furthermore, in spite of the latter fact it is always $f_i(\text{Li}) < f_i(\text{Na}) < f_i(\text{K}) < f_i(\text{Rb})$. Then, if we assume that the bulk modulus — volume — ionicity relationship for the alkali halides is analogous to that for the tetrahedral semiconductors (NEUMANN) it should be expected that the exponent n has the same value for all compounds while the factor B_0 in (1) is constant for compounds with the same cation but differs for compounds with different cations.

3. Analysis of bulk modulus data

In analysing the bulk modulus-volume relationship in the alkali halides we first consider the isothermal bulk modulus B^T. A complete list of isothermal compressibilities for room temperature has been given by TOSI, isothermal elastic constants for all compounds have been compiled by GARG et al. The resulting isothermal bulk moduli are given in Table 2. The accuracy of these values is certainly not better than ±3% if we consider the scatter in the literature data for the elastic constants of the alkali halides (see e.g. LEWIS et al.; LOIDL et al.; SHAHAM et al.; SLAGLE, MCKINSTRY). We note that within this error limit the isothermal bulk moduli reported by other authors (BARTELS, SCHUELE; CHANG, BARSCH; MCLEAN, SMITH; MILLER, SMITH; ROBERTS, SMITH 1970a, b; YAMAMOTO et al.) agree with those given in Table 2.

Table 2
Comparison of experimental and calculated isothermal bulk moduli (room temperature) of the alkali halides with rocksalt structure

Compound	B^T_{exp}/GPa		B^T_{calc}/GPa
	TOSI	GARG et al.	
LiF	67.1	66.40	68.65
LiCl	29.8	29.59	29.63
LiBr	23.8	23.56	23.45
LiI	17.2	17.27	17.28
NaF	46.5	46.55	47.49
NaCl	24.0	23.71	23.96
NaBr	19.9	19.45	19.67
NaI	15.1	14.91	14.92
KF	30.5	30.22	30.49
KCl	17.5	17.35	17.38
KBr	14.8	14.64	14.76
KI	11.7	11.62	11.66
RbF	26.3	26.68	26.35
RbCl	15.6	15.55	15.33
RbBr	13.0	13.23	13.16
RbI	10.5	10.48	10.57

Figure 1 shows a double-logarithmic plot of B^T against the unit cell volume $V_0 = a^3$. From this plot it is evident that it is impossible to describe the bulk moduli of all compounds by means of relation (1) using the same values for n and B_0 for the whole compound family as it has been supposed by MITRA and MARSHALL and by SIRDESHMUKH and SUBHADRA, even if the given error limit for the experimental data is taken into account. On the other hand, if we follow JAYARAMAN et al. and consider compounds with the same cation separately, it is obvious from Figure 1 that relation (1) with a constant value of B_0 gives an excellent description of the bulk modulus-volume dependence. Moreover, the slopes of the curves we find for compounds with different cations is nearly the same in all cases. In other words, the exponent n of relation (1) seems to be the same for all alkali halides, whereas the factor B_0 is a constant only for compounds with the same cation or, as it can be seen from Table 1, with the same bond ionicity.

To quantify the parameters n and B_0 of relation (1) we start with the data for the potassium halides. Since they all have the same ionicity (Tab. 1) their bulk moduli can be used to determine the exponent n with sufficient reliability. A least-squares fit of relation (1) to the experimental B^T values for the potassium halides gives $n = 1.147$. Then, we assumed that the exponent n is the same for all compounds and that B_0 is constant for compounds with the same cation (see Fig. 1) and determined the corresponding values for B_0^T. The resulting B_0^T values are compiled in Table 3,

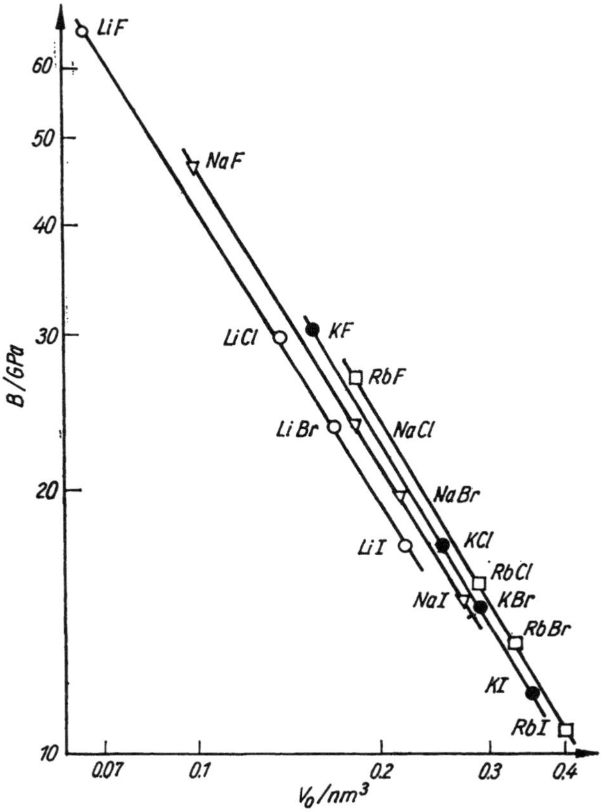

Fig. 1. Double-logarithmic plot of the isothermal bulk modulus against the unit cell volume for the alkali halides with rocksalt structure

Table 3
Coefficients B_0^T of relation (1) for $n = 1.147$ (if B^T is inserted in GPa and V_0 in nm^3)

Compounds	f_i	f_c	B_0^T
Li halides	0.904	0.096	3.00
Na halides	0.934	0.066	3.34
K halides	0.953	0.047	3.53
Rb halides	0.957	0.043	3.65

the bulk moduli B^T calculated with these B_0^T values and $n = 1.147$ are given in the last column of Table 2. We see that within the supposed error limit of $\pm 3\%$ the calculated B^T values agree with the experimental ones. The largest deviation between the calculated and experimental bulk modulus we have for LiF (see Tab. 2), but in this compound the scatter in the literature data for the elastic constants is relatively large (SHAHAM et al.) and it is quite possible that the true B^T value for LiF is somewhat larger than that proposed by TOSI and GARG et al.

Now, let us consider the ionicity dependence of B_0^T. From Table 3 it follows that B_0^T increases with increasing ionicity f_i or decreasing covalency $f_c = 1 - f_i$ of the chemical bond. In other words, at the same unit cell volume a compound with $f_i = 1$ (or $f_c = 0$) has the largest bulk modulus. This tendency in the dependence of the bulk modulus on the bond ionicity is opposite to that observed in the tetrahedral semiconductors where a decrease of B_0 with increasing ionicity or loss of covalency has been found (COHEN; NEUMANN). The dependence of B_0^T on f_i or f_c can be well described by a linear function. Using f_c as the critical parameter we can write

$$B_0^T = b_0(1 - b_1 f_c) \tag{3}$$

with $b_0 = 4.134$ and $b_1 = 2.877$. Relation (3) reproduces the empirically determined B_0^T values with an accuracy of better than $\pm 2\%$.

Finally, some remarks regarding the scaling of the adiabatic bulk modulus B^S which is usually determined in experimental measurements. At a given temperature T the relation between the adiabatic and isothermal bulk modulus is given by

$$B^S = B^T \left\{1 + \frac{\alpha^2 V_m B^T T}{C_V}\right\}, \tag{4}$$

where α is the volume expansion coefficient, V_m the molar volume and C_V the molar heat capacity at constant volume. From relation (4) it is obvious that at sufficiently high temperatures the second term in the brackets becomes important and the scaling law for B^S is expected to be different from that for B^T. Comparing the B^S values for room temperature (see e.g. HAUSSÜHL) with the corresponding B^T values of Table 2 we find that for LiF and all sodium, potassium and rubidium halides we have $B^S = (1.05 \pm 0.01)\ B^T$. This means that for these compounds B^S scales again according to relation (1) with $n = 1.147$ and $B_0^S \approx 1.05 B^T$. In LiCl, LiBr, and LiI the ratio B^S/B^T is equal to 1.071, 1.085 and 1.110, respectively, which deviates already from the value of about 1.05 for the other compounds.

References

BARTELS, R. A., SCHUELE, D. E.: J. Phys. Chem. Solids **26** (1965) 537
BROUT, R.: Phys. Rev. **113** (1959) 43
CHANG, Z. P., BARSCH, G. R.: J. Phys. Chem. Solids **32** (1971) 27
COHEN, M. L.: Phys. Rev. B **32** (1985) 7988
GARG, V. K., PURI, D. S., VERMA, M. P.: phys. stat. sol. (b) **80** (1977) 63
HAUSSÜHL, S.: Z. Physik **159** (1960) 223
JAYARAMAN, A., BATLOGG, B., MAINES, R. G., BACH, H.: Phys. Rev. B **26** (1982) 3347
LANDOLT-BÖRNSTEIN: Numerical Data and Functional Relationships in Science and Technology (New Series), Vol. III/7a, Berlin-Heidelberg-New York 1973
LEVINE, B. F.: J. Chem. Phys. **59** (1973) 1463
LEWIS, J. T., LEHOCZKY, A., BRISCOE, C. V.: Phys. Rev. **161** (1967) 877
LOIDL, A., DAUBERT, J., MÜLLNER, M.: Z. Physik B **30** (1978) 235
MAJEWSKI, J. A., VOGL, P.: Phys. Rev. B **35** (1987) 9666
McLEAN, K. O., SMITH, C. S.: J. Phys. Chem. Solids **33** (1972) 279
MILLER, R. A., SMITH, C. S.: J. Phys. Chem. Solids **25** (1964) 1279
MITRA, S. S., MARSHALL, R.: J. Chem. Phys. **41** (1964) 3158
NEUMANN, H.: Cryst. Res. Technol. **22** (1987) 99
ROBERTS, R. W., SMITH, C. S.: J. Phys. Chem. Solids **31** (1970a) 619
ROBERTS, R. W., SMITH, C. S.: J. Phys. Chem. Solids **31** (1970b) 2397
SHAHAM, J., KAPLAN, H., LOW, W., FOGUEL, M.: Phys. Rev. Lett. **24** (1970) 82
SIRDESHMUKH, D. B., SUBHADRA, K. G.: J. Appl. Physics **59** (1986) 276
SLAGLE, O. D., McKINSTRY, H. A.: J. Appl. Physics **38** (1967) 437
SRIVASTAVA, V. K.: Phys. Rev. B **29** (1984) 6993

SZIGETI, B.: Proc. Roy. Soc. A **204** (1950) 52
TOSI, M. P.: Solid State Physics **16** (1964) 1
VAN VECHTEN, J. A.: Phys. Rev. **187** (1969) 1007
WEMPLE, S. H.: J. Chem. Phys. **67** (1977) 2151
YAMAMOTO, S., OHNO, I., ANDERSON, O. L.: J. Phys.Chem. Solids **48** (1987) 143

(Received, accepted October 6, 1987)

Authors' address:

Dr. H. NEUMANN
Sektion Physik, Karl-Marx-Universität
Linnéstraße 5
Leipzig
DDR-7010

Cryst. Res. Technol. | 23 | 1988 | 4 | 537–542

A. M. RADWAN

Theoretical Physics Laboratory, National Research Centre, Cairo, Egypt

Madelung Constants of the Superlattices of the Laves Phases

The Madelung constants of all possible superlattices of the Laves phases are calculated using a modified form of the Ewald-Fuchs method.

Die Madelungkonstanten aller unterschiedlicher Überstrukturen der Lavesphasen werden mit Hilfe einer modifizierten Ewald-Fuchs-Methode berechnet.

1. Introduction

A modified form of the Ewald-Fuchs method (FUCHS) is used to evaluate the Madelung constants of the superlattices of the Laves phases C14, C15 and C36 (SCHULZE; PEARSON). The structures considered can be expressed as a stacking sequence of three different types of layers in a way similar to the analysis given by KOMURA. In the expression for the Madelung constant (HEINE, WEAIRE) the structure factor is then written as a product of

a) a structure factor of the individual layers which is structure independent within the considered structures and
b) a structure factor of the layer sequences which is structure dependent and the reciprocal space sum is also appropriately carried out.

The real space sum is eliminated through suitable choice of the radius of the Gaussian caps. Our method is equivalent to that of plane-wise summations as discussed by SHOLL.

2. Description of the structures

All the structures considered have the form of a binary compound XY_2. They are described in terms of the stacking variants of the following three different types of layers:
1. simple close-packed layers containing atoms of the sort X, with three possible layer positions α, β, and γ,
2. simple close-packed layers of some Y-atoms (YI) with the layer positions a, b and c,
3. Kagomé layers of the remaining atoms of type Y (YII). The three possible layer positions here are denoted by A, B, and C.

The planes α, β, γ, a, b, c, and A, B, and C differ in the position of their inversion centre. For the α, a, A planes it lies at the origin, for β, b, B at $\boldsymbol{a}_1/3 + \boldsymbol{a}_2/3$ and for γ, c, C at $\boldsymbol{a}_1/3 - \boldsymbol{a}_2/3$ where $\boldsymbol{a}_1$ and $\boldsymbol{a}_2$ are the primitive vectors of a two-dimensional hexagonal lattice. Figure 1 shows the base of the unit cell with the different layer positions.

Each simple layer contains one atom in the unit cell but there are three atoms of the type YII in each Kagomé sheet. The ideal value for the parameter x in the Kagomé sheets B and C is $x = 1/6$. For the height z of the X-atoms above the planes α, β, and γ,

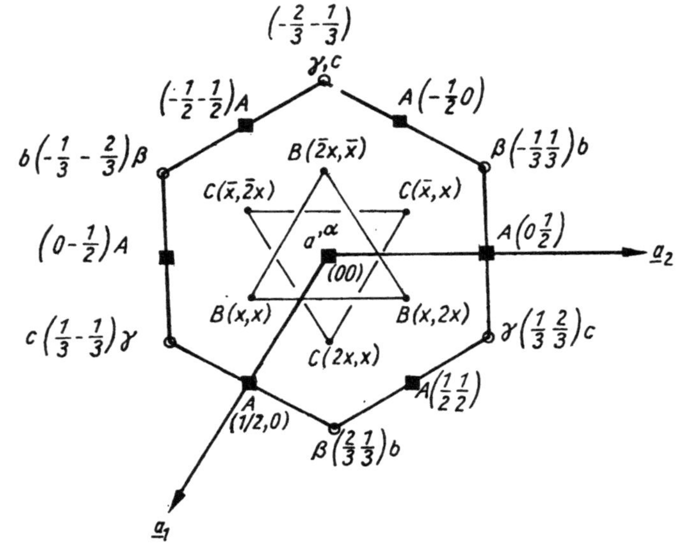

Fig. 1. The base of the unit cell showing the different layer positions

the ideal value is $z = 1/16$. In terms of the layers described above the structure of the Laves phases C14, C15 and C36 can be expressed as:

C14 ($MgZn_2$) γCγaβBβa = CB
C15 ($MgCu_2$) αAαcβBβaγCγb = ABC (only with ideal x, z, and c_0/a_0)
C36 ($MgNi_2$) αAαcβBβcαAαbγCγb = ABAC.

A, B, and C on the right-hand side characterize the stacking packages αAαb or αAαc, βBβc or βBβa, and γCγa or γCγb, respectively, each being of height $c_0/2$ and containing six atoms. The unit cells of the structures C14, C15, and C36 have then the heights c_0, $1.5c_0$, and $2c_0$, respectively.

Table 1
Perpendicular distance between neighbouring layers (in units $c_0/2$)

Layer types	distance	example
Kagomè-Kagomé	1	A—B, B—C, C—A
Kagomé-simple	3/8	A—α, β—B, C—γ
simple-simple	1/8	a—γ, α—b, β—c

It should be noticed that in the Kagomé sheet A the parameter x always has its ideal value (PEARSON). Table 1 contains the perpendicular distance between neighbouring layers in the structures under consideration with ideal value of parameter z.

3. Method of calculation

In the Ewald-Fuchs method the Madelung constant α_{es} is given by (HEINE, WEAIRE):

$$\alpha_{es} = 3R_a^{-2} \sum_{g}' |S(\boldsymbol{g})|^2 g^{-2} e^{-g^2/4\eta} - (2R_a \sqrt{\eta/\pi} + 3R_a^{-2}\eta^{-1}/4) + + R_a \sum_{r_i}' |\boldsymbol{r}_i|^{-1} G(\sqrt{\eta}\, |\boldsymbol{r}_i|) \,. \tag{3.1}$$

In this formula $\boldsymbol{g}$ denotes non-zero reciprocal lattice vectors of structure factor $S(\boldsymbol{g})$, $\boldsymbol{r}_i$ is a real lattice vector, G is the complementary error function, and R_a is related to the volume per atom V by $V = 4\pi R_a^3/3$. η is an arbitrary parameter determining the radius of the Gaussian caps and hence the rate of convergence of the two series in reciprocal and real space. We chose a relatively large η to eliminate the real space sum since the band structure energy is also most conveniently evaluated in reciprocal space and the summation can be performed in the same way. The Madelung constant Eq. (3.1) can then be written as

$$\alpha_{es}(J) = E_g(J) + C' , \tag{3.2}$$

where $E_g(J)$ is structure dependent and C' is a constant. The electrostatic energy per ion E_{es} is related to α_{es} by:

$$E_{es} = Z^2\alpha_{es}/2R_a , \tag{3.3}$$

where Z is the valence.

For a binary alloy with N atoms in the unit cell the structure factor $S(\boldsymbol{g})$ is given by:

$$S^J(\boldsymbol{g}) = \sum_{M=1}^{2} S_M^J(\boldsymbol{g}) \tag{3.4}$$

where

$$S_M^J(\boldsymbol{g}) = \frac{1}{N} \sum_{j}^{\text{M-atoms}} e^{-i\boldsymbol{g}\cdot\boldsymbol{r}_j^{(M,J)}} \tag{3.5}$$

is the structure factor for the atoms of type M and structure J. The summation is carried out only over atoms of type M in the unit cell. $E_g(J)$ has then the form

$$E_g(J) = \sum_{\boldsymbol{g}}' \sum_{M,M'} S_M^J(\boldsymbol{g})\, S_{M'}^{*J}(\boldsymbol{g})\, f(g) . \tag{3.6}$$

where

$$f(g) = 3R_a^{-2} g^{-2} e^{-g^2/4\eta} . \tag{3.7}$$

Through splitting of $\boldsymbol{r}_j$ in Eq. (3.5) into $\boldsymbol{r}_j = \boldsymbol{r}^{\|} + \boldsymbol{r}^{\perp}$ where $\boldsymbol{r}^{\|}$ is parallel to the layers α, a, A etc. and $\boldsymbol{r}^{\perp}$ is perpendicular to them and doing the corresponding splitting in $\boldsymbol{g}$, $\boldsymbol{g} = \boldsymbol{g}^{\|} + \boldsymbol{g}^{\perp}$, Eq. (3.5) becomes

$$S_M^J(\boldsymbol{g}) = \frac{1}{N} \sum_{L=1}^{3} S1(L, \boldsymbol{g}_{\|})\, S2(M, L, \boldsymbol{g}_{\|})\, SM(J, L, \boldsymbol{g}_{\perp}) . \tag{3.8}$$

L characterizes the layer position. $S1(L, \boldsymbol{g}_{\|})$ is the structure factor of the inversion centre of the layer with position L. $S2(M, L, \boldsymbol{g}_{\|})$ is the structure factor of the layer in the position L and containing atoms of type M.

$SM(J, L, \boldsymbol{g}^{\perp})$ is the structure factor of the sequence of layers with position L and atom type M. It should be noticed that both structure factors $S1(L, \boldsymbol{g}_{\|})$ and $S2(M, L, \boldsymbol{g}_{\|})$ are independent on the structure under consideration, but $SM(J, L, \boldsymbol{g}^{\perp})$ is a structure dependent one.

Substituting for $S_M^J(\boldsymbol{g})$ from Eq. (3.8) into Eq. (3.6) and after some modifications we get

$$E_g(J) = \frac{1}{N^2} Re \sum_{\boldsymbol{g}_{\perp}} \sum_{M=1}^{3} \sum_{M'=M}^{3} \sum_{L=1}^{3} \sum_{L'=K}^{3} (2 - \delta_{MM'}\, \delta_{LL'})\, SM(J, L, \boldsymbol{g}_{\perp}) \times \times SM^*(J, L', \boldsymbol{g}_{\perp})\, W_{MM'LL'}(\boldsymbol{g}_{\perp}) , \tag{3.9}$$

where in the summation over L': K = L if M = M' and K = 1 otherwise and

$$W_{MM'LL'}(\boldsymbol{g}_{\perp}) = \sum_{\boldsymbol{g}_{\|}} S1(L, \boldsymbol{g}_{\|})\, S^*1(L', \boldsymbol{g}_{\|})\, S2(M, L, \boldsymbol{g}_{\|})\, S^*2(M', L', \boldsymbol{g}_{\|}) \times \times f(\sqrt{\boldsymbol{g}_{\|}^2 + \boldsymbol{g}_{\perp}^2}) . \tag{3.10}$$

4. Results

We have used Eq. (3.9) to calculate Madelung constants of the superlattices of the Laves phases consisting of all the different possible stacking variants up to ten Kagomé nets given in Table 2.

Table 2
All the possible and non-equivalent stacking variants (superlattices) up to ten Kagomé sheets with ideal x

Number of layers	structure	sequence	number of layers	structure	sequence
2	C14	CB	9	9c	BCABACABC
3	C15	ABC	9	9d	BCACABABC
4	C36	ABAC	9	9e	ABACABABC
5	5	BCABC	9	9f	ABABACABC
6	6a	BABCAC	9	9g	ABABABCAC
6	6b	ABACAC	10	10a	ABCBCACBCB
7	7a	CBCABCB	10	10b	BCBABCACBC
7	7b	CABACAB	10	10c	BCBACBACBC
7	7c	CACBABA	10	10d	ABCABACABC
8	8a	ABABABAC	10	10e	ABCACABABC
8	8b	ABABACAC	10	10f	BABABACABA
8	8c	BCABCABC	10	10g	BABABACACA
8	8d	ABCACBAB	10	10h	BABACABACA
8	8e	CBCABCBA	10	10i	BABABACBCA
8	8f	BCACBABC	10	10j	CABACABCBA
9	9a	BCBCABCBC	10	10k	BACACABABC
9	9b	BCACBABCA	10	10l	BABACABCAC

Table 3
Madelung constants of the superlattices of the Laves phases up to ten Kagomé sheets calculated with ideal values of x, c_0/a_0, and z, arranged in ascending order of $|\alpha_{es}|$

Structure	$-\alpha_{es}$	structure	$-\alpha_{es}$	structure	$-\alpha_{es}$
C14.	1.771514	10k	1.771841	5	1.771987
10f	1.771694	10a	1.771841	10l	1.771987
10g	1.771694	10h	1.771841	10j	1.771987
8b	1.771731	8d	1.771914	6a	1.772035
8a	1.771731	8f	1.771914	8c	1.772096
9a	1.771791	C36	1.771914	8e	1.772096
9g	1.771792	9f	1.771954	9b	1.772116
9d	1.771792	9e	1.771954	10d	1.772133
6b	1.771792	9c	1.771954	C15	1.772312
10i	1.771840	10c	1.771986		
10b	1.771841	10e	1.771986		

The Madelung constants, calculated for ideal values of c_0/a_0, x and z, are given in Table 3 in ascending order.

From the 31 structures given in Table 3, the Madelung constant of only three structures: C14, C15, and C36 has been calculated before by HAFNER. α(C14) was also computed by FUKS et al. This gives us the possibility of comparing our results with those of other authors to judge our method. The values of α quoted by HAFNER are:

$$\alpha_{es}(C14) = -1.771515 ,$$

$$\alpha_{es}(C\bar{3}6) = -1.771920 ,$$

$$\alpha_{es}(C15) = -1.772313 ,$$

which are in perfect agreement with our results. On the other side, we believe that $\alpha_{es}(C14) = -1.64166$ of FUKS et al. is wrong. For the other 28 structures of Table 3 there are no published values of α_{es} in the literature. Thus, our results represent the first calculation of the Madelung constants of these 28 superlattices.

Now, we want to study the effect of changing the parameter x of the Kagomé nets B and C on the values of the Madelung constants of the different structures at ideal c_0/a_0 and z. It should be noticed that the Kagomé net A is always taken with $x_{ideal} = 1/6$. This means that the three planes A, B, and C are equivalent as long as x has its ideal value. However, if x is changed then both planes B and C will remain equivalent to each other, but not to plane A. Consequently, we have, e.g., AB = AC but AB $\neq$ CB if $x \neq 1/6$. This will increase the number of non-equivalent superlattices of Table 2 considerably. Thus, for practical reasons, we confine ourselves to the structures of Table 2. We have taken $x = 0.17$, which is the experimental value of C14 (PEARSON) and calculated α_{es} for all structures of Table 3 with ideal c_0/a_0 and z. The results are given in Table 4 in ascending order.

Table 4
Madelung constants of the superlattices of Table 3 calculated with $x = 0.17$ and ideal c_0/a_0 and z, arranged in ascending order of $|\alpha_{es}|$

Structure	$-\alpha_{es}$	structure	$-\alpha_{es}$	structure	$-\alpha_{es}$
10g	1.770140	10k	1.771017	8f	1.772170
10f	1.770140	10e	1.771175	8c	1.772371
8b	1.770180	10l	1.771176	8e	1.772372
8a	1.770180	10j	1.771178	10b	1.772450
6b	1.770247	8d	1.771273	10a	1.772450
10h	1.770302	10d	1.771337	10c	1.772610
C36	1.770382	9d	1.771441	5	1.772610
9g	1.770644	9c	1.771620	9a	1.773035
9f	1.770823	6a	1.771705	C14	1.773566
9e	1.770823	9b	1.771794		
10i	1.771017	C15	1.771973		

Comparing Tables 3 and 4, we see that changing x from x_{ideal} to $x = 0.17$ has a remarkable influence on α_{es}:

1. For example in the ten-layers structures $|\alpha_{es}|$ increases for those having few A planes (10a, 10b, 10c) and decreases as the number of A planes increases (10f, 10g).
2. The minimum of $\alpha_{es}(C14)$ is at the experimental value $x = 0.17$ and not the ideal x.
3. C15 structure in Table 4 is, strictly speaking, a distorted structure since in the ideal C15, $x = 1/6$. This is demonstrated clearly in the calculated $\alpha_{es}(C15)$, which has its minimum at x_{ideal} rather than $x = 0.17$, because of symmetry reasons.
4. $\alpha_{es}(C36)$ has its minimum at x_{ideal}, which is the experimental value (PEARSON).

References

FUCHS, K.: Proc. Roy. Soc. A **151** (1935) 585
FUKS, D. L., ZHOROVKOV, M. F., PANIN, V. E.: Sov. Phys. J. (USA) **18** (1975) 1357
HAFNER, J.: J. Phys. F **6** (1976) 1243
HEINE, V., WEAIRE, D.: Solid State Physics **24**, New York 1970
KOMURA, Y.: Acta Cryst. **15** (1962) 770
PEARSON, B. W.: A Handbook of Lattice Spacings and Structures of Metals and Alloys, Vol. 2, New York 1967
SCHULZE, G. E. R.: Metallphysik, Berlin 1967
SHOLL, C. A.: Proc. Phys. Soc. **92** (1967) 434

(Received September 24, 1987; accepted October 2, 1987)

Author's address:

Dr. A. M. RADWAN
National Research Centre
Theoretical Physics Laboratory
Tahrir Street
Cairo-Dokki
AR Egypt

Cryst. Res. Technol.	23	1988	4	543–548

J. BISCHOF, K. BOHMHAMMEL, P. DEUS

Research Institute of Electrical Engineering, Prague, and
Mining Academy, Freiberg

Specific Heat and Related Properties of Some $A^{II}B^{IV}C_2^V$ and $A^IB^{III}C_2^{VI}$ Semiconducting Compounds between 2 and 300 K

From specific heat measurements of the ternary semiconducting compounds $ZnSiP_2$, $ZnSiAs_2$, $CdGeAs_2$, $CuGaSe_2$, and $CuInTe_2$ Debye characteristic temperatures and standard entropy values are calculated within the temperature range from 2 to 300 K.

Aus Messungen der spezifischen Wärmekapazität der ternären halbleitenden Verbindungen $ZnSiP_2$, $ZnSiAs_2$, $CdGeP_2$, $CuGaSe_2$, und $CuInTe_2$ werden Debyetemperaturen und Standardentropien im Temperaturbereich von 2–300 K berechnet.

1. Introduction

The complicated chalcopyrite structure of the ternary semiconducting compounds $A^IB^{III}C_2^{VI}$ and $A^{II}B^{IV}C_2^V$ results into more pronounced effects of anisotropy and anharmonicity than in their binary sphalerite analoga $A^{II}B^{VI}$ and $A^{III}B^V$.

Because of the lack of monocrystalline samples of ternary compounds suitable in size and shape the main part of experimental work in lattice dynamics of these substances has been done by means of classical methods such as thermal expansion, specific heat, mean square amplitudes of lattice oscillations etc. Specific heat measurements, for instance, yield informations on temperature dependence of Debye characteristic temperature, and allow estimations of phonon frequencies, interatomic force constants, or lattice anharmonicity.

LEROUX HUGON and VEYSSIE measured the specific heat of $CdSnAs_2$ and $ZnSnAs_2$ below 4.2 K and calculated Debye characteristic temperatures $\theta_0 = 234$ K and 271 K, resp., at 0 K. ABRAHAMS and HSU derived from measurements within 1.2 ... 40 K the following θ_0: $ZnGeP_2$ – 428 K, $CuGaS_2$ – 356 K, $CdGeP_2$ – 340 K, and $AgGaS_2$ – 255 K. BACHMANN et al. evaluated temperature dependences of specific heat and Debye characteristic temperature of $CuInSe_2$ ($\theta_0 = 222$ K), $CuInTe_2$ ($\theta_0 = 191$ K), and $AgGaTe_2$ ($\theta_0 = 182$ K) within 1 ... 40 K, and $CuInS_2$ ($\theta_0 = 273$ K) and $AgInTe_2$ ($\theta_0 = 156$ K) within 1 ... 300 K. From the measurements up to 300 K DEUS et al. estimated interatomic force constants by means of a simplified Keating model. BOHMHAMMEL et al. calculated Debye and Einstein characteristic temperatures of $ZnSiP_2$ ($\theta_0 = 445$ K), $ZnSiAs_2$ ($\theta_0 = 347$ K), $ZnSnAs_2$ ($\theta_0 = 266$ K), $CdGeAs_2$ ($\theta_0 = 241$ K), $CuGaSe_2$ ($\theta_0 = 262$ K), $CuGaTe_2$ ($\theta_0 = 226$ K), and $CuInTe_2$ ($\theta_0 = 195$ K) from specific heat measurements within 2 ... 30 K and gave an estimate on shape and frequencies of acoustic phonon modes of these compounds. Recently, NEUMANN et al. and MÖLLER et al. derived from specific heat measurements within a range between 300 and 500 K data on lattice anharmonicity in the compounds $ZnSiP_2$, $ZnSiAs_2$, $CdB^{IV}C_2^V$, $AgInS_2$, $AgGaTe_2$, and $CuAlTe_2$. Generally, the main part of specific heat measurements of the ternary compounds mentioned has been done at very low temperatures or above 300 K up to the moment. It is the aim of this work to measure specific heat of some $A^{II}B^{IV}C_2^V$ and $A^IB^{III}C_2^{VI}$ compounds within a temperature range of 2 ... 300 K in order to fill a gap in C_p-measurements of ternary chalcopyrite compounds and to calculate temperature dependence of Debye characteristic temperature and further thermodynamical quantities within this temperature range.

2. Experimental

The used equipment suitable for experimental study of heat properties of materials in the temperature range from 1.5 to 300 K was developed in the Research Institute of Electrical Engineering (VÚSE) in Prague, Czechoslovakia, where the measurements were carried out. The equipment basically consists of a bath cryostat and of a bath insert cryostatcalorimeter, into which the measured sample can be put (Bischof).

A modified classical Nernst method was used for the experimental evaluation of heat capacity of the sample. Here the sample after having been quasi-adiabatically separated from the bath is heated by pulses.

The lowest temperature of the sample is reached by filling the cryostat with gaseous helium at a partial pressure of approximately 1 Pa. After the required temperature is reached, the chamber with the sample is pumped off for a long period, until a steady state with respect to helium desorption from the sample surface is reached.

The sample is surrounded by an isothermal shield, which is kept at a constant temperature near the temperature of the sample. The effect of parasitic heat flows between shield and sample reducing the precision of measurement was compensated by continuously measuring the temperature of the sample between the heat pulses. The effect of the parasitic heat capacity (temperature sensor, heater, adhesives, cover of sample) was eliminated by measuring this capacity in a special experiment.

The specific heat of the sample is defined by equ. (1)

$$C_p(T) = \frac{\Delta Q}{m \cdot \Delta T}, \tag{1}$$

where ΔQ is the heat added to the sample, m is the mass of the sample, and ΔT is the increase of the sample temperature. This temperature is estimated by means of extrapolation of the time dependences of the sample temperature before and after the heat pulses. The temperature is extrapolated to the time centre of each pulse.

The very diluted acetone adhesive DUOSAN was mixed with the powdered ternary compounds (the mass of the samples was between 1.4 to 15 g) and then the samples were pressed into a thin-walled copper-nickel tube. The electric manganin heater was fixed to the tube by means of varnish GE 7031. This varnish was applied for the fixing of two calibrated resistance temperature sensors as well, — a germanium sensor for the temperature range from 1.5 – 100 K and a platinum sensor for 60 – 300 K. The electric

Table 1
Relative errors of specific heat measurements in %

sample	temperature range in K			
	4–20	20–40	40–60	60–300
$ZnSiAs_2$	3	4	6	3
$CdGeAs_2$	3	5	8	4
$ZnSiP_2$	2	2	3	2
$CuGaSe_2$	2	3	3	2
$CuInTe_2$	2	2	3	2

resistance of the sensors was measured in four-point arrangement by means of a special resistance bridge.

The reliability of the apparatus described here was tested by measuring a sample of very pure copper (99.995%). The results of this test comply with the results of representative measurements given by FURUKAWA et al.

The $A^{II}B^{IV}C_2^{V}$ samples used in the experiments were grown from the vapour phase. The $A^{I}B^{III}C_2^{VI}$ compounds were prepared by direct fusion of stoichiometric mixtures of the elements. Structure and lattice parameters of the substances under investigation were evaluated by X-ray diffraction. The relative error of the C_p-measurements which depends on the sample mass available and on the temperature range is given in Table 1.

3. Results

The dependences of specific heat at constant pressure C_p on temperature for $ZnSiP_2$, $ZnSiAs_2$, $CdGeAs_2$, $CuGaSe_2$, and $CuInTe_2$ are given in Figure 1. It should be noted that the C_p-values refer to a number of $6.022 \cdot 10^{23}$ atoms, here, i.e. the values given for molecules e.g. by NEUMANN et al. have to be divided by four for comparison.

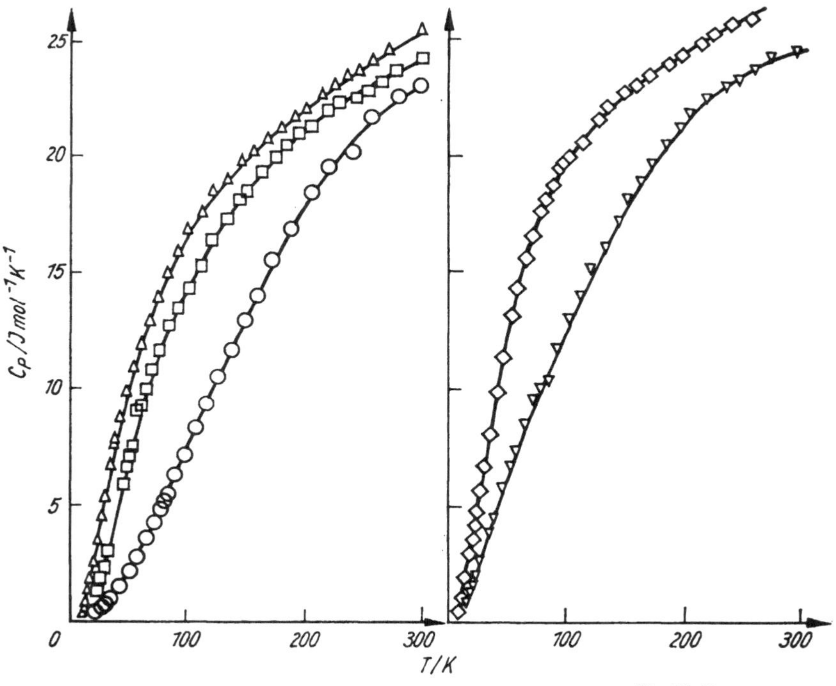

Fig. 1. Specific heat C_p at constant pressure vs temperature; left: $A^{II}B^{IV}C_2^{V}$ – compounds, right: $A^{I}B^{III}C_2^{VI}$ – compounds. ○ $ZnSiP_2$, □ $ZnSiAs_2$, △ $CdGeAs_2$, ▽ $CuGaSe_2$,C ◇ $CuInTe_2$

As expected, in both groups the compounds with weaker binding forces or lower lattice frequencies, respectively — which generally corresponds to a larger molecular mass — have the higher specific heat values at a certain temperature. Near 300 K there is a small deviation from the values given by NEUMANN et al. and MÖLLER et al. which amounts to 4% for $CdGeAs_2$, 6% for $ZnSiP_2$, and 7% for $ZnSiAs_2$, respectively. This is of the same order as the experimental errors evaluated and should be explained by the different types of calorimeters used which are specialized for low or high temperature measurements, respectively.

From the specific heat values the Debye characteristic temperatures θ_D were calculated according to

$$C_v = 3R\left\{\frac{4\pi^4}{5x^3} - \sum_{n=1}^{\infty}\left[3nx + 12 + \frac{36}{nx} + \frac{72}{(nx)^2} + \frac{72}{(nx)^3}\right]\frac{e^{-nx}}{n}\right\}, \tag{2}$$

for $T < 40$ K and according to

$$C_v = 3R\left(1 - \frac{x^2}{20} + \frac{x^4}{560} - \frac{x^6}{18144}\right), \tag{3}$$

for $T > 40$ K. Here, R is the gas constant and $x = \theta_D/T$. The difference between C_P and C_v has not been taken into account because its maximum value is of the same order as the experimental error within the temperature range investigated.

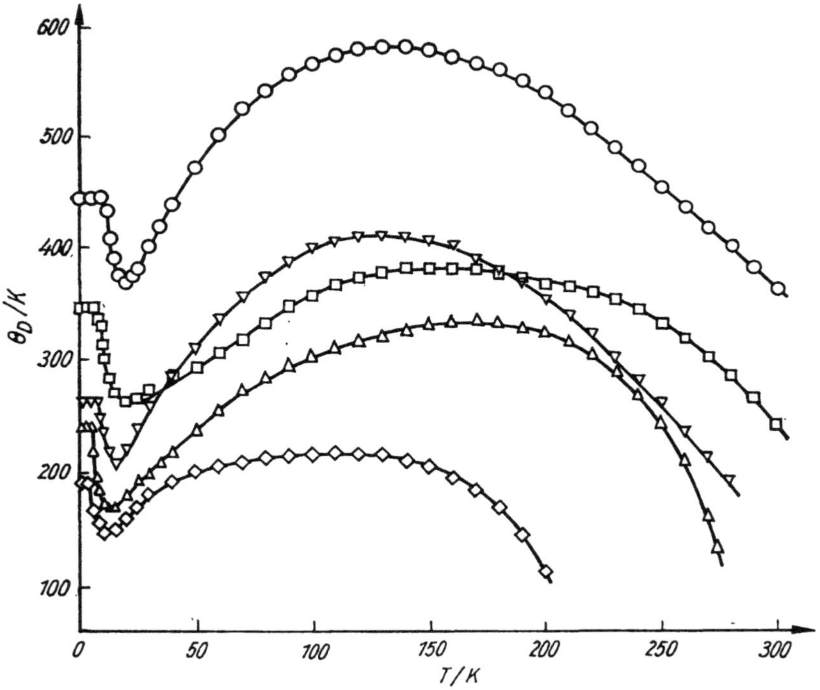

Fig. 2. Debye characteristic temperature θ_D vs temperature (symbols as in Figure 1)

The values of θ_D and the shape of the low temperature part of the temperature dependences of θ_D (Fig. 2) are in accordance with independent measurements of Bohmhammel et al. and are represented and discussed in detail in the publications named.

In Figure 3 the relative Debye characteristic temperatures θ_D/θ_0 are compared. From this comparison follows that the degree of anharmonicity which is responsible for the decrease of Debye characteristic temperatures at high temperatures, differs to some extent in the compounds under investigation. Thus the fact found by Neumann

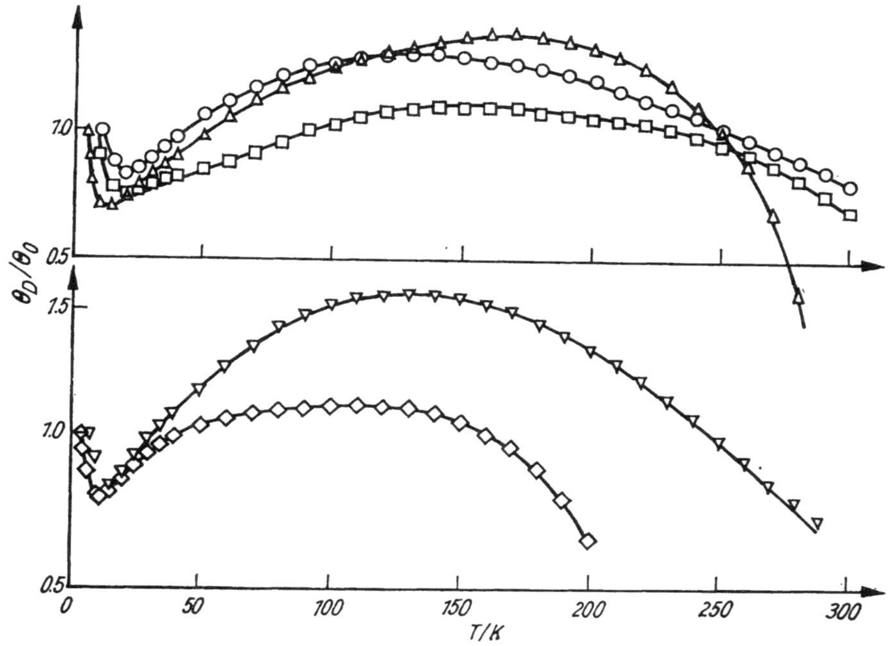

Fig. 3. θ_D/θ_0 vs temperature (symbols as in Figure 1)

et al., that the lattice anharmonicity in $ZnSiP_2$ is more pronounced than in $ZnSiAs_2$, is confirmed by the corresponding curves in Figure 3.

Finally, the standard entropy of the investigated compounds which may be used for further thermodynamical considerations were calculated by means of

$$S^{\theta}(T) = \int_0^T C_P \, d(\ln T) \, . \tag{4}$$

The results at various temperatures are given in Table 2. The mean relative error of the entropy values is $\approx \pm 2\%$.

Table 2
Entropy S^{θ} of the investigated substances in J Mol^{-1} K^{-1}

T/K	100	200	298
$ZnSiP_2$	4.06	12.61	20.90
$ZnSiAs_2$	10.49	22.80	32.26
$CdGeAs_2$	15.45	28.83	38.25
$CuGaSe_2$	9.97	21.77	31.06
$CuInTe_2$	19.51	34.97	45.19

The authors are grateful to Dr. H. Neumann (University of Leipzig) and Dr. V. Frei (University of Prague) for the discussion of some aspects of this work, to Prof. Dr. G. Kühn (University of Leipzig) and Dr. E. Buhrig (Bergakademie Freiberg) for providing of the substances used in the present investigation, and to P. Hollstein for the technical assistance.

References

Abrahams, S. C., Hsu, F. S. L.: J. Chem. Phys. **63** (1975) 1162
Bachmann, K. J., Hsu, F. S. L., Thiel, F. A., Kasper, H. M.: J. Electron. Mater. **6** (1977) 431
Bischof, J.: Czech. J. Phys. **A 29** (1979) 599
Bohmhammel, K., Deus, P., Schneider, H. A.: phys. stat. sol. (a) **65** (1981) 563
Bohmhammel, K., Deus, P., Kühn, G. Möller, W.: phys. stat. sol. (a) **71** (1982) 505
Deus, P., Frei, V., Bilek, O., Cordts, W.: phys. stat. sol. (b) **112** (1982) 269
Deus, P., Frei, V.: Czech. J. Phys. **B 34** (1984) 680
Furukawa, G. T., Saba, W. G., Reilly, M. L.: Critical analysis of the heat-capacity data of the literature and evaluation of thermodynamic properties of copper, silver, and gold from 0 to 300 K, NBS Washington, NSR DS — NBS 18. April 1968
Leroux Hugon, P., Veyssie, J. J.: phys. stat. sol. 8 (1965) 561
Neumann, H., Kühn, G., Möller, W.: Cryst. Res. Technol. **20** (1985) 1225
Neumann, H., Kühn, G., Möller, W.: Cryst. Res. Technol. **21** (1986) 919
Möller, W., Kühn, G., Neumann, H.: Cryst. Res. Technol. **22** (1987) 533 K18, K57

(Received September 30, 1987;
accepted October 8, 1987)

Authors' addresses:

Dr. J. Bischof
Research Institute of Electrical Engineering
25097 Praha 911 — Běchovice, Czechoslovakia

Dr. sc. nat. K. Bohmhammel
Bergakademie Freiberg, Sektion Chemie
Leipziger Straße
Freiberg
DDR-9200

Dr. sc. nat. P. Deus
Bergakademie Freiberg, Sektion Physik
Cottastraße 4
Freiberg
DDR-9200

Cryst. Res. Technol. | 23 | 1988 | 4 | 549–554

Ji. Kvapil, B. Mánek, B. Perner, Jos. Kvapil, R. Becker, G. Ringel

Monokrystaly, Turnov, Czechoslovakia, and
Leybold-Heraeus, Köln, Bundesrepublik Deutschland

The Role of Argon in Yttrium Aluminates

Spectral properties of YAG and YAP grown in vacuum or under Ar–H_2 protective atmosphere were compared. The latter contained small amounts of argon which was detected by a laser mass spectral analysis. Ar impurity evokes brown colour of unannealed crystals and strenghtens anomalous red colour of annealed (particularly in O_2) YAP crystals doped with Nd + Cr or Nd + Mo. The same red coloration of the doped YAGs is less stable.

Es werden spektrale Eigenschaften der im Vakuum oder unter Ar–H_2 Atmosphäre gezüchteten Yttrium-Aluminate verglichen. Die Einkristalle enthalten eine kleine Menge Argon, die man mittels Laser Massen-Spektralanalyse nachweisen kann. Eine braune Färbung der ungetemperten YAG- und YAP-Einkristalle ebenso wie die anomale rote Färbung der mit Nd + Cr oder mit Nd + Mo dotierten und getemperten (besonders in O_2) YAP-Einkristalle werden durch die Anwesenheit von Ar erklärt. Die letztgenannte rote Färbung der dotierten YAG-Kristalle ist relativ unstabil.

1. Introduction

Gain and losses of the solid state laser crystals grown from Y_2O_3–Al_2O_3 system depend on the existence of various point defects to which belong the presence of impurity metallic ions (Holloway, Kestigian; Kvapil et al. (a); Powell, Merkle; Riseberg; Williams) and colour centres of different stability due to impurities or nonstochiometry (Bass, Paladino; Bernhardt; Kvapil et al. (b); Mori). Colour centres are associated usually with defects of oxygen sublattice as oxygen vacancies or interstitials (e.g. Holloway, Kestigian; Pacheco, di Bartolo; Weber) or as the holes localized in the neighbourhood of large cations on the cationic sites of a relatively small dimension (Kvapil et al. (c)).

The latter may be divided into three groups:

a) The centres due to the presence of Mg^{2+} after oxygen annealing (yellow colour) seem to be relatively independent of preparation method.
b) Brown colour of unannealed crystals rich of Y^{3+} is typical of that grown in non-oxidizing conditions, particularly, in Ar–H_2 atmosphere.
c) Relatively stable red colour of annealed YAP:Nd codoped with Cr or Mo and transient one of YAG:Nd were found in the crystals grown under Ar–H_2 atmosphere.

The aim of this paper is to prove the presence of argon in YAG and YAP crystals even after annealing in vacuum or other gases and to explain their spectral properties.

2. Experimental

YAG:Nd and YAP:Nd crystals containing ~ 1 at.% Nd/Y were grown by Czochralski method using molybdenum crucibles, resistance heating and 95% Ar + 5% H_2 protective atmosphere (Perner et al.) or vacuum. The atmosphere was dried (dew point −30 °C) to avoid any dopage of the crystals with Mo ions (KVAPIL et al. (d)). The crystal slices and rods up to 7 mm in thickness (diameter) were alternatively annealed in H_2, O_2 or vacuum, respectively. 1700 °C and 3 h are quite sufficient to avoid brownish coloration due to growth conditions. Some samples were annealed for an extremely long time (200 h) at the same temperature in $H_2O_{(g)}$ or in the mixture of 90–95% O_2 + 5–10% J_2 (DEVOR et al.).

Dopant and impurity concentration in the crystals were determined by X-ray fluorescence and neutron activation analysis, but Ar as well as light elements may be detected using a LAMMA-1000 (Leybold Heraeus) laser mass spectral analyzer (HEINEN et al.) only. Optical absorption spectra, luminescence lifetime, transient absorption and laser active parameters were measured using the methods described earlier (KVAPIL et al. (a)). Some crystals were melted down in vacuum and the gas escaped from the furnace was excited using gold electrodes (distance: 1.2 cm, power supply: 15 kV 2 MΩ) in a chamber with quartz window. The spectrum was analyzed using a Zeiss SPM-2 monochromator and a photomultiplier.

3. Results

Strong anomalous (brown and blue) colouration of the unannealed crystals is proportionate to the Ar content in the protective atmosphere. A weak colouration of vacuum grown crystals is in accordance with this observation (KVAPIL et al. (b)). On the other hand, the admixture of some transient metal ions, e.g. Ti, suppresses such colouration even in the presence of argon. Surprisingly, Ar content (Fig. 1) in the

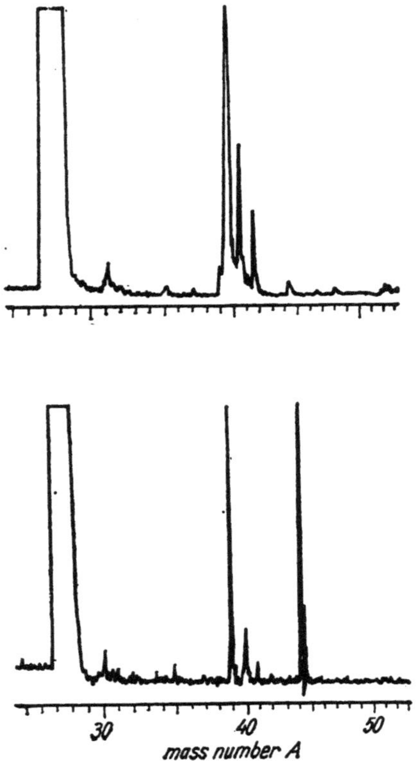

Fig. 1. Mass spectrum of YAP grown in Ar–H_2 atmosphere (above) and after prolonged annealing in $H_2O(g)$ (below) with peaks belonging to the mass number (A): 27 (Al), 40 (Ar), and 44 (AlOH)

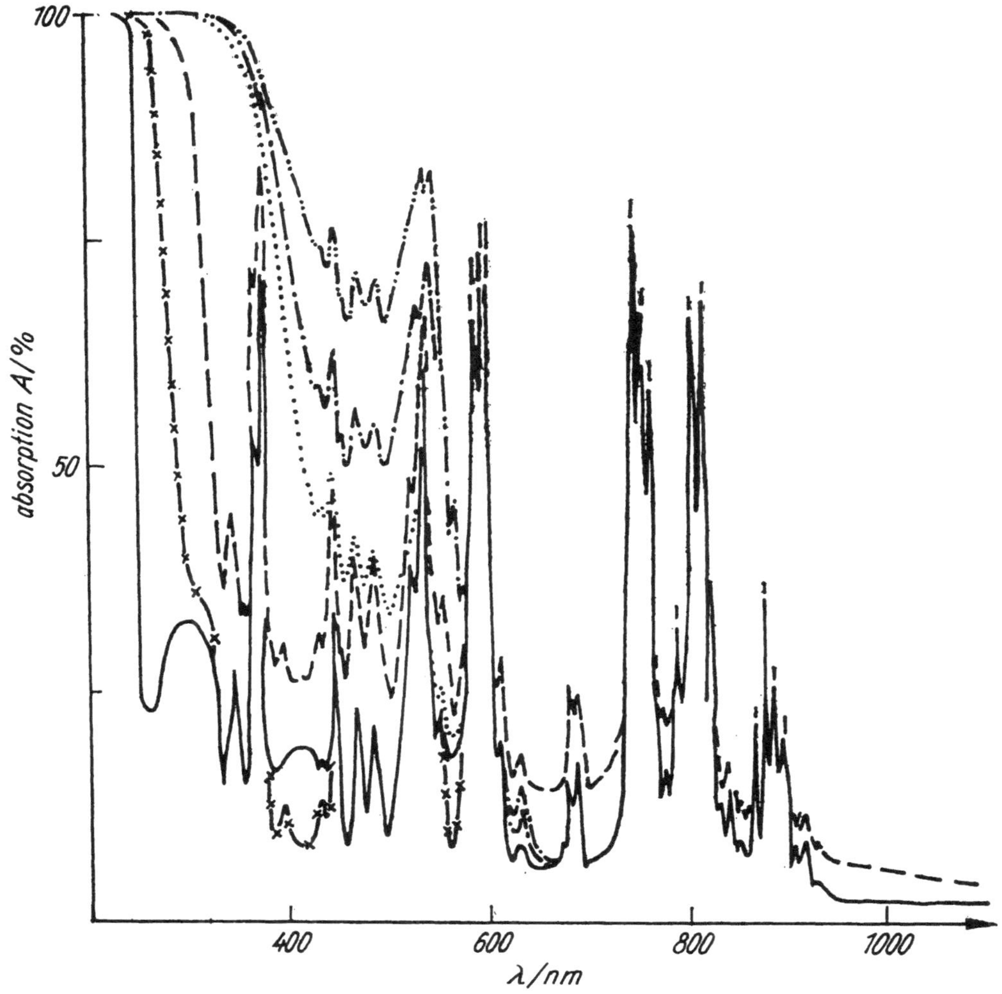

Fig. 2. Absorption (A) of YAP:Nd (0.8 wt% Nd) grown in Ar–H_2: 1 (— — —) unannealed; 2 (—•—•—) codoped with $3 \cdot 10^{-3}$ wt% Cr annealed in O_2 (approximately the same absorption showed the same samples annealed in H_2 after Xe-lamp flashing or that annealed in O_2–J_2 for 150 h); 3 (—••—••—) sample 2 after 50 flashes of Xe-lamp (100 J); 4 (•••••) codoped with $3 \cdot 10^{-3}$ wt% Cr + $3 \cdot 10^{-3}$ wt% Ce, annealed in O_2, after flashing; 5 (————) codoped with $3 \cdot 10^{-3}$ wt% Cr + $3 \cdot 10^{-2}$ wt% Ce + $3 \cdot 10^{-4}$ wt% Fe annealed in O_2, after flashing. 6 (— × — × —) absorption of YAP:Nd, Ce free of Cr + Fe grown and annealed in vacuum

crystals grown in Ar–H_2 atmosphere was independent on the annealing in O_2, H_2, or vacuum. It was slightly lowered after prolonged (> 150 h) annealing in the O_2–J_2 mixture or $H_2O_{(g)}$ only. Vacuum melting of the crystals grown under Ar–H_2 atmosphere causes that all the Ar content is quickly released. This was detected during several minutes after the crystals had been melted down as U_2 (811.53 nm) and U_4 (750.39 nm) lines (Saidel et al.) Ar in the annealed Nd-doped crystals conditiones red colouration (absorption between 300–650 nm), particularly after oxygen treatment and/or UV irradiation.

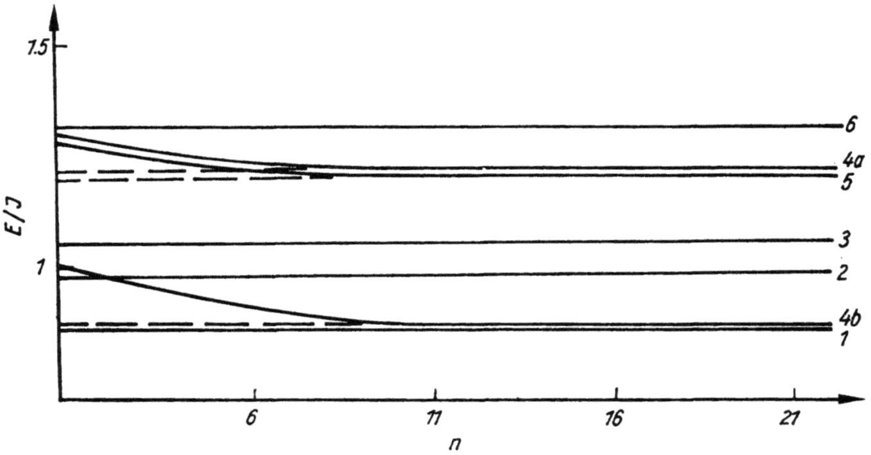

Fig. 3. Dependence of the pulsed output of 1 – YAG:Nd; 2 – YAG:Nd, Ce, Cr ($1 \cdot 10^{-2}$ wt% Ce, $3 \cdot 10^{-2}$ wt% Cr); 3 – YAP:Nd, 4 – YAP:Nd, Cr ($8 \cdot 10^{-3}$ wt% Cr); 5 – YAP:Nd, Ce, Cr ($1 \cdot 10^{-2}$ wt.% Ce, $3 \cdot 10^{-3}$ wt% Cr) and 6 – YAP:Nd, Ce, Cr, Fe ($1 \cdot 10^{-2}$ wt% Ce, $3 \cdot 10^{-3}$ wt% Cr, $3 \cdot 10^{-4}$ wt% Fe) on the shot number (n). The rods of 6 mm in diameter and 75 mm in length were machined of the crystals grown under Ar–H_2 atmosphere and annealed in H_2 (1, 2, 3, 4 a, 5, 6) or O_2 (4 b). Ag coated single elliptical cavity and filter with absorption edge at 300 mm was used. Sample 6 gave the same results using no filter. Input 100 J, resonator length: 400 mm, R_{out}: 30 %. Full line: the first application, dashed line: following applications

The colouration is usually relatively stable in YAP:Nd and less stable (transient after UV-irradiation) in YAG:Nd (KVAPIL et al. (a)). An extremely strong colouration of this type, which is dependent on the Ar content in the growth atmosphere, was observed in YAP:Nd, Cr and/or YAP:Nd, Cr, Fe (10^{-3} wt% Cr or Fe) but never in YAP: Nd, Cr, Ce containing even a small concentration of Fe ($\gtrapprox 2 \cdot 10^{-4}$ wt%) (Figs. 2, 3).

It is a substantial difference to YAP: Nd, Mo, Fe (KVAPIL et al. (d)), where reduction treatment should be used to avoid such a coloration. The crystals codoped with Cr^{3+} (up to 0.05 wt% Cr) were also successfully grown under the protective atmosphere only. Chromium oxides are volatile in the oxidizing atmosphere and extremely volatile in vacuum. The crystals containing Fe were grown immediately after the raw material containing this dopant had been melted down because of reduction of this dopant. Nevertheless, the crystals containing up to 2–3 ppm Fe could be grown 40–80 h after dopage. Vacuum grown YAP crystals cannot be doped with Cr as well as with Mo. Anomalous red coloration was not observed in this case.

4. Discussion

High temperature ionization of argon causes some portion of argon ions to be dissolved in the growing crystal. During cooling, Ar^+ ions serve as hole donors. The hole may turn "compressed" O^{2-} ions in $[Y_{Al}O_6]$ octahedrons into O^-. It is typical of brown colour of unannealed crystals grown under Ar–H_2 atmosphere from the melt rich of Y_2O_3 (KVAPIL et al. (b)). Vacuum grown crystals showed a weak colouration because of the absence of Ar. Annealing at high temperature causes the superfluous cations to be removed and Y_{Al}^{3+} enter regular, i.e. dodecahedral site where no compression of O^{2-} in its neighbourhood is possible. Nevertheless, Ar at high temperature is ionized again and bound in the lattice. That is why the removing of Ar from the crystal is very difficult. It may be partially accomplished by the annealing in the other ionized gas ($H_2O_{(g)}$ or O_2–J_2 mixture).

The hole released from Ar^+ after thermal treatment may react with F centre to form a relatively stable anionic vacancy. After an intense irradiation in UV absorption bands of Nd^{3+} serves this vacancy as an electron acceptor from $[NdO_8]$ dodecahedron and O^- (bound polaron) in the neighbourhood of Nd^{3+} with absorption between 300 to 650 nm (red colour) is formed. It is transient in the case of YAG:Nd and stable in the case of YAP:Nd codoped with Cr^{3+} particularly in oxygen treated samples. The role of Cr^{3+} in both types of the crystals may be explained on the basis of different dimensions of octahedral sites. Small amount of Cr^{3+} ($< 10^{-3}$ wt%) in YAG enter relatively large octahedral sites only and may serve as a hole trap only (instead of $[NdO_8]$) to form Cr^{4+}_{Al} and no anomalous red colouration is observed. On the other hand using a greater concentration of Cr^{3+} in YAG:Nd and even traces of Cr^{3+} in YAP:Nd (with relatively small octahedral sites) (KVAPIL et al. (c)), chromium ions enter also large dodecahedral sites where its further reduction is possible.

There is an equilibrium given by the reaction:

$$Cr^{3+}_Y + [NdO_8] \rightleftarrows Cr^{2+}_Y + [NdO_8^{(+)}] ,$$

which is shifted to the right side by UV irradiation. It means that Cr^{3+}_Y is a hole donor. The ratio of $Cr^{4+}_{Al}:Cr^{3+}_{Al}$ as well an $Cr^{3+}_Y:Cr^{2+}_Y$ is directly proportionate to the Ar concentration and oxygen/metal ratio in the crystal.

The latter is increased (decreased) after oxygen (vacuum and especially hydrogen) treatment. Using more efficient hole acceptor, e.g. Ce^{3+}, the $Cr^{3+}_Y:Cr^{2+}_Y$ ratio as well as $[NdO_8^{(+)}]$ concentration is decreased. The charge transfer from Cr^{3+}_Y or $[NdO_8^{(+)}]$ to the Ce^{3+} seems to be facilitated by the presence of iron in a mixed valence state even in the concentration of $2-3 \cdot 10^{-4}$ wt%. The presence of both Fe^{3+} and Fe^{2+} is typical of annealed YAP (KVAPIL et al. (a)) where the above mentioned concentration of iron together with 10—1000 time greater concentration of cerium prevent the formation of the "red colour centre" (absorption between 250 and 650 nm). On the other hand annealed YAGs contain iron as trivalent ions only and the charge transfer from Cr^{3+}_Y or $[NdO_8^{(+)}]$ is influenced by the presence of iron to a relatively small extent. Enlarged concentration of iron ($\gtrapprox 10^{-3}$ wt%) causes always undesirable effects as luminescence quenching and increased absorption of the crystals exposed to an UV irradiation. Nevertheless, the YAP:Nd crystals containing Ce, Cr and up to 10^{-3} wt% Fe showed an efficient $Cr^{3+} \rightarrow Nd^{3+}$ energy transfer without detrimental effect of colour centre and they may be used as an excellent active laser material. It is to be noted, that optically perfect YAP crystals containing Cr were grown under Ar–H_2–(He) protective atmosphere only.

5. Summary

YAG and YAP crystals grown under Ar–H_2 protective atmosphere contain dissolved Ar which may be detected by laser mass spectral analysis or in the gaseous species escaped during the remelting of the crystals. Ar forms at high temperature positive ions which loses during cooling its positive charge. The released holes react with F centre, $[Y_{Al}O_6]$ dodecahedron and with some transitive or rare earth ions, e.g. Cr^{2+}_Y, Cr^{3+}_{Al} or Ce^{3+}. The resulting $[Y_{Al}O_6^{(+)}]$, Cr^{4+}_{Al} and particularly Ce^{4+} seem to be relatively stable, whereas anionic vacancy and particularly Cr^{3+}_Y may serve as electron trap from $[NdO_8]$ dodecahedron which form $[NdO_8^{(+)}]$. The charge transfer between the hole centres and Ce^{3+} is facilitated by iron in a mixed valence state which is typical of annealed YAPs. Brown colour of unannealed crystals may be attributed to $[Y_{Al}O_6^{(+)}]$ and red colour of Nd-doped crystal to $[NdO_8^{(+)}]$ centre. Vacuum grown crystals showed a limited brown colouration before annealing and no anomalous red colour (Nd-doped crystals) after annealing because of the absence of argon.

References

BASS, H., PALADINO, A. E.: J. Appl. Physics **38** (1967) 2706

BERNHARDT, H.-J.: phys. stat. sol. (a) **21** (1974) 95; (a) **37** (1976) 445; (b) **87** (1978) 213

DEVOR, D. P., PASTOR, R. C., DE SHAZER, L. G.: J. Chem. Phys. **81** (1984) 4104

HEINEN, H. J., MEIER, S., VOGT, H., WECHSUNG, R.: Internat. J. Mass Spectrometry and Ion Physics **47** (1983) 19

HOLLOWAY, W. W., KESTIGIAN, M.: J. Opt. Soc. Amer. **59** (1969) 60

KVAPIL, J., KVAPIL, JOS., KUBELKA, J., PERNER, B.: Czech. J. Phys. **B 32** (1982) 817; KVAPIL, JOS., KVAPIL, J., KUBELKA, F., AUTRATA, R.: Cryst. Res. Technol. **18** (1983) (a) 127

KVAPIL, J., KVAPIL, JOS., PERNER, B., MÁNEK, B., BLAŽEK, K., HENDRICH, Z.: Cryst. Res. Technol. **20** (1985) (b) 473

KVAPIL, J., KVAPIL, JOS., PERNER, B., MÁNEK, B.: Kristall und Technik **10** (1975) 529; KVAPIL, J., KVAPIL, JOS., KUBELKA, J., KUBEČEK, V.: Czech. J. Phys. **B 31** (1981) 644; KVAPIL, J., KVAPIL, JOS., PERNER, B., MÁNEK, B., BLAŽEK, K.: Cryst. Res. Technol. **21** (1986) (c) 349

KVAPIL, J., PERNER, B., KVAPIL, JOS., KUBELKA, J., HAMAL, K., KOŠELJA, M.: Czech. J. Physics **B 36** (1986) (d) 751

MORI, K.: phys. stat. sol. (a) **42** (1977) 375

PACHECO, D., DI BARTOLO, B.: J. Lumin. **16** (1978) 1

PERNER, B., KVAPIL, J., KVAPIL, JOS.: Czech. J. Physics **B 24** (1974) 1091

POWELL, R. C., MERKLE, L. D.: Phys. Rev. **B 20** (1979) 75

RISEBERG, L. A.: in: Radiationless Processes, (eds B. DI BARTOLLO and V. GOLBERG), New York 1980, p. 369

SAIDEL, A. N., PROKOFJEW, W. K., RAJSKI, S. M.: Spektral-Tabellen, Berlin 1955, p. 539

(Received September 30, 1987;
accepted October 8, 1987)

Authors' addresses:

Dr. Jiři KVAPIL, B. MÁNEK, B. PERNER, JOS. KVAPIL
Monokrystaly
51119 Turnov, ČSSR

Dr. Gerhard RINGEL, Rosemarie BECKER
Leybold-Heraeus GMBH
Bonner Straße 498
D-5000 Köln, FRG

Cryst. Res. Technol.	23	1988	4	555–560

K. S. CHANDRA BABU, O. N. SRIVASTAVA

Physics Department, Banaras Hindu University, Varanasi, India, and
Department of Chemistry, Banaras Hindu University, Varanasi, India

Photoelectrochemical Solar Cells Based on $TiS_2/TiO_2(TiO_xS_y)$ Photoelectrodes

A new photoelectrode of TiS_2/TiO_2 alloy with a possible new TiO_xS_y phase has been synthesized for the first time in the present investigation. This photoanode exhibits improved photoelectrochemical (PEC) response as compared to either TiS_2 or TiO_2 photoelectrode. With this new photoelectrode high PEC conversion efficiency is evidenced by large photovoltage (840 mv) and significantly high photocurrent density (40 mA/cm^2). The photovoltage obtained in the present system is significantly high and the photocurrent density is the highest reported so far for the TiO_2 based systems. It has been suggested that the enhancement in photoactivity in the present system is due to the reduced band gap of TiO_xS_y phase in TiS_2/TiO_2 alloy material. Such a band gap reduction is thought to be facilitated by the downward movement of the valence band of titanium disulphide which is made up of chalcogen atoms through substitution of sulphur atoms by oxygen atoms.

Introduction

In view of the possible exhaustion of energy resources in near future, coupled with the need for avoiding pollution, efficient conversion of solar energy into useful forms including electrical and chemical variants is one of the most important problems of current research. Hydrogen is an excellent substitute for oil and has the added advantage of having nearly zero pollution. The photoelectrochemical (PEC) conversion of solar energy into electricity has several attractive features over its all solid state version. As for example, the production of fuel (hydrogen) and viable storage are features novel to PEC route (PRASAD et al.). The PEC solar cells suffer from the perenial problem of photocorrosion. Some of the oxide photoelectrodes (e.g. TiO_2, $SrTiO_3$, etc.) are comparatively less susceptible to photocorrosion and are, therefore, stable. However, due to large band gap (~ 3 eV) they exhibit poor solar spectral response. On the other hand the low band gap semiconductors like Si, CdS, GaAs, CdSe etc. ($E_g \sim 1$ eV to ~ 2.2 eV) show good spectral response but undergo rapid photocorrosion under PEC conditions. There is an online search for efficient photoelectrode materials with good stability and better spectral response. Among the large number of oxides explored, large band gap oxides are found to be stable, but with very small solar spectral absorptivity, whereas the low band gap oxides exhibit either poor stability or positive flat band potential both of which are detrimental in regard to PEC conversion. After the discovery of FUJISHIMA and HONDA in 1972 that photoelectrolysis of water can be achieved via PEC route through TiO_2 photoanode many materials as photoelectrodes have been investigated. Recently it appears that in this paradoxical material problem, the interest has shifted back to the oxide materials, especially towards TiO_2 (HARTIG et al.; WRIGHTON).

Several attempts have been made to bring the spectral response of TiO_2 into visible region. Doping of some transition metal ions into TiO_2 was able to bring some shift in its spectral response towards visible region (GHOSH, MARUSKA; SALVADOR; HOULIHAN et al.). Various mixed oxides with TiO_2 as one ingredient have also been tried for this purpose, since it is a known fact that the catalytic activity of the binary oxide is often higher than that of their individual components. Titanium oxide has been able to show enhanced photocatalytic activity when alloyed with SiO_2 at a particular composition (MASAKAZU ANPO et al.). NiO, Co_3O_4., and Fe_2O_3 have also been tried in doping concentration with TiO_2 for their enhanced photoactivity (ZIELINSKI, SOBCZYNSKI). (Ti, V)O_2- and (Ti, Nb)O_2-type of alloys have also been experimented over for their improved PEC activity (GAUTRON et al. 1983). TiO_2–VO_2 alloys with the composition $Ti_{1-x}V_xO_2$ ($0.04 \leqq x \leqq 0.55$) have been produced with reduced band gap. In this binary oxide the reduction of band gap has been obtained by the downward movement of conduction band of TiO_2, which is made up of d-orbitals of the transition metal. This type of movement being disadvantageous for PEC conversion (PHILLIPS et al.). Another possible mode of the lowering of band gap of TiO_2 would be to substitute chalcogen atoms like S/Se for the oxygen atoms. This is expected to lead to the upward movement of valence band of TiO_2 and hence reduction in band gap. In the present investigation this mode of band gap reduction has been attempted. In the substitution of oxygen atoms by sulphur atoms to various extents, the incorporation of sulphur into TiO_2 single crystal could not be achieved due to higher affinity of oxygen towards titanium. Therefore, in the present investigation single crystals of titanium disulphide have been subjected to controlled oxidation at different temperatures, facilitating controlled replacement of sulphur with oxygen, leading in effect to the formation of TiO_xS_y-type of phase, presumably having a lower band gap than TiO_2. The material TiS_2 is an MX_2-type layered semiconductor with an indirect p–d band gap of 0.3 ± 0.2 eV (MURRAY, YOFFE).

Experimental techniques, results, and discussion

Single crystals of titanium disulphide have been grown by iodine vapour transport technique (CHITRA PANDE et al.). Pure elements (99.99%) of titanium and sulphur in stoichiometric proportions were placed in an evacuated quartz tube and homogenized at 473 K for 24 h. This powder was transferred into another evacuated quartz growth tube with iodine transporter. A vapour transport in a two-zone furnace with reaction zone at 1073 K and growth zone at 993 K was carried out for about 400 h. Platelet-shaped crystals a few mm across and having thicknesses of about 15 μm were found to grow in the growth zone. The titanium disulphide single crystals so obtained were subjected to controlled annealing in oxygen atmosphere in a quartz tube. The controlled oxidation was performed at three different temperatures of 1073, 873, and 723 K followed by annealing in hydrogen atmosphere at their corresponding oxidation temperatures. The hydrogen reduction in the present case is expected to lead to S and O vacancies. This in keeping with the known results (SUBBA RAO et al.; HARTIG et al.) would result in an increased conductivity. The crystals oxidized/reduced at 1073, 873, and 723 K would be referred to as the first, second, and third category of crystals, respectively. The greyish crystals so obtained were subjected to structural and electrochemical characterisation.

Structural characterisation

The as oxidized/reduced crystals were found to be polycrystalline by X-ray diffraction technique. The X-ray diffractograms obtained from Philips PW-1710 diffractometer have confirmed the presence of titanium dioxide, titanium sulphide and some other phases. A representative X-ray diffractogram of the oxidized/reduced TiS_2 corresponding to the third category of crystals is shown in Figure 5. A detailed analysis of the diffractograms revealed the presence of TiO_2 (rutile), TiS, and Ti_3S_4 phases and several other lines from an unknown phase. Such lines have been marked 'P' in the diffractogram. A rigorous analysis in regard to the indexing of these lines revealed that they did not belong to any known structural system in the Ti–O and Ti–S phases. In the light of this, it may presumably be taken that these lines correspond to TiO_xS_y-typ ephase. Such a phase would possess a lower band gap than TiO_2, since the replacement of sulphur atoms with oxygen would cause the downward movement of the valence band of TiS_2.

Photoelectrochemical characterisation

PEC measurements were carried out in a pyrex glass cell with a quartz window. The counter electrode was a Pt foil ($\sim 2\ cm^2$) and Saturated Calomel Electrode (SCE) the reference electrode. A Princeton Applied Research (PAR) 175 Universal Programmer and PAR 173 Potentiostat/Galvanostat were used to monitor the potential at the electrode and the corresponding current obtained through PAR 173 I/E converter was recorded on Houston Model 2000 X – Y recorder. The electrode was prepared by pasting silver on one side of the crystal which is fixed over Pt, sealed in pyrex glass. The edges of the crystal were sealed from electrolyte with epoxy resin. A 1000W Xe–Hg lamp (Oriel Corporation) was used for electrode illumination, through a water filter reducing heating effects. All the current-voltage (I – V) measurements were performed in a standard buffer of *p*H 13.00. Solutions were made from analytical grade chemicals and doubly distilled water.

As mentioned in literature (LEITH, TARSHALL), titanium disulphide starts getting oxidized at 673 K and gets completely transformed to titanium dioxide (rutile) at 1073 – 1273 K. In view of this the oxidation followed by reduction was carried out at three different temperatures, enabling the controlled oxidation of titanium disulphide. As revealed by the X-ray diffractograms, the major components of these samples constituted of titanium dioxide and titanium sulphide together with a new phase of type TiO_xS_y. In view of the possible lower band gap of TiO_xS_y phase as compared to TiO_2 and consequent photocurrent enhancement, the I – V curves were recorded for all the three categories of crystals, both under dark and illuminated conditions. These curves have been compared with those of TiO_2 and TiS_2 for their improved PEC activity.

Figure 1 shows I – V characteristics of the $TiS_2(A_D, A_L)$ and TiO_2 (B_D, B_L) single crystals, both under dark (D) and illumination (L). As is evident, titanium disulphide shows hardly any PEC activity, whereas TiO_2 exhibits reasonable activity as expected. Figure 2 exhibits I – V characteristics bringing out the PEC activity of the first category of TiS_2 crystals and Figure 3 represents corresponding curves of the second category of crystals. Figure 4 brings out the PEC activity of the third category of crystals.

From a comparison of Figures 2 and 3, it can be seen that the crystals corresponding to the second category show a considerable enhancement of PEC activity. As for instance, the photopotential and photocurrent corresponding to the first category of crystals are 640 mV and 4 mA/cm^2 whereas for the second category of crystals these are 770 mV and 15 mA/cm^2. The same for the third category of crystals are 840 mV and 40 mA/cm^2. The first and second category of crystals oxidized/reduced at higher temperatures (1073 and 873 K) apparently have a higher density of titanium dioxide than the third category of crystals (oxidized/reduced at 723 K). This is confirmed by their X-ray diffractograms. The spectral and PEC response of the first and second

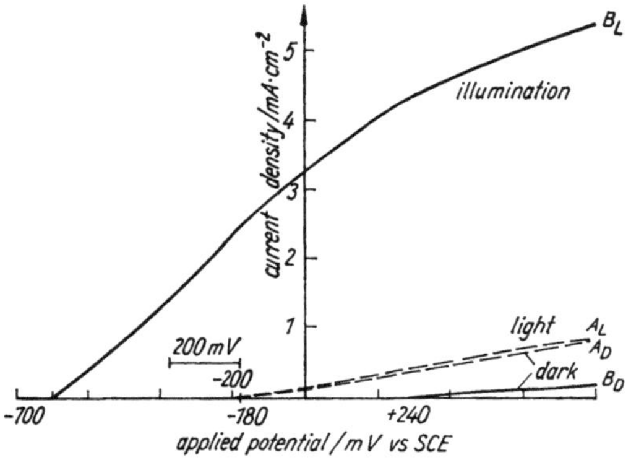

Fig. 1. Current-voltage curves of TiS_2 and TiO_2 single crystals under dark (A_D, B_D) and illumination (A_L, B_L) in a buffer of pH 13.00

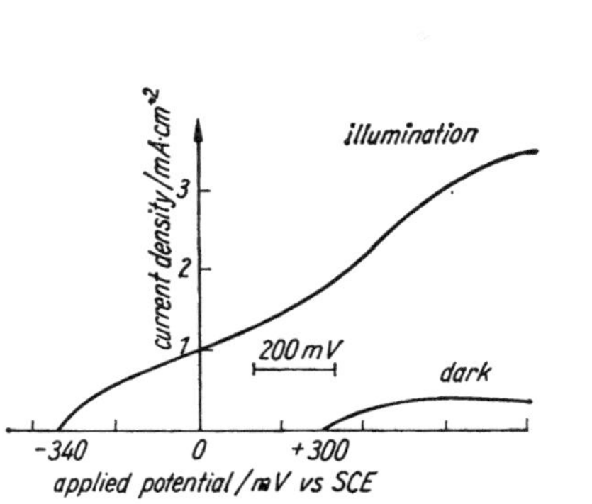

Fig. 2. Current-voltage curves of first category of crystals (oxidized/reduced at 1073 K) under dark and illumination in a buffer of pH 13.00

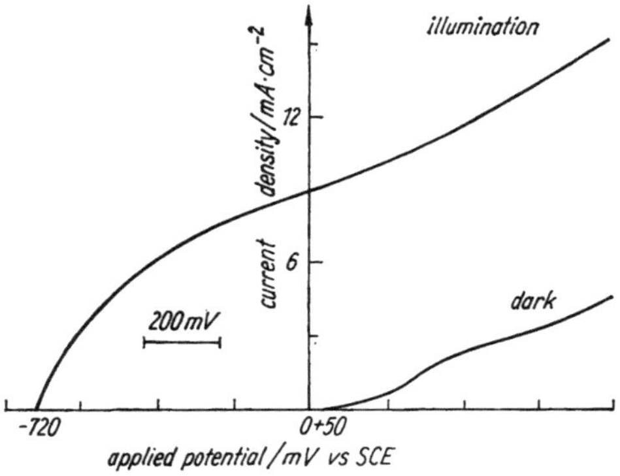

Fig. 3. Current-voltage curves of second category of crystals (oxidized/reduced at 873 K) under dark and illumination in a buffer of pH 13.00

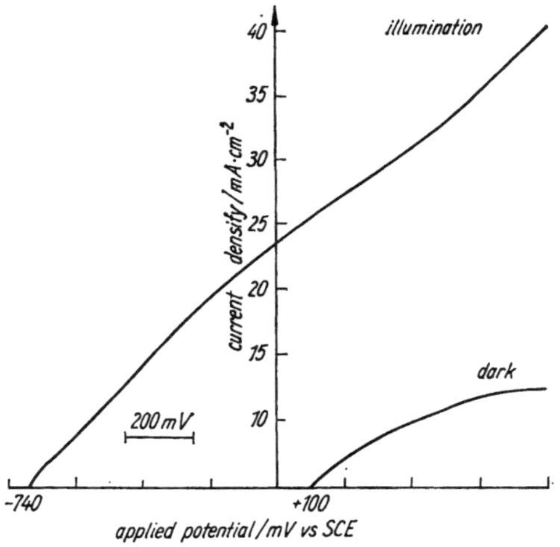

Fig. 4. Current-voltage curves of third category of crystals (oxidized/reduced at 723 K) under dark and illumination in a buffer of pH 13.00

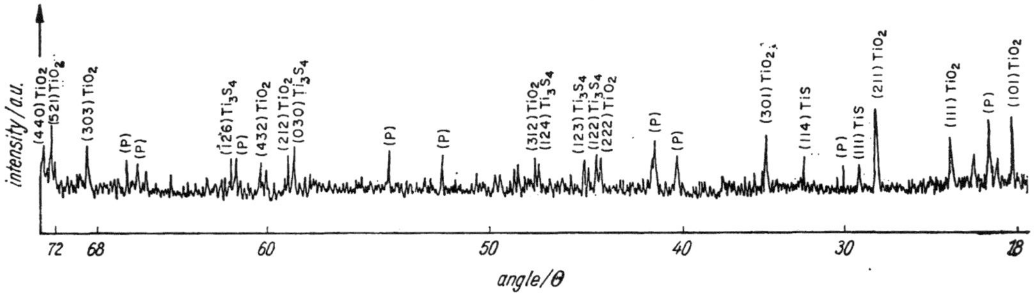

Fig. 5. Typical X-ray diffractogram of the third category of crystals exhibiting TiO_2, TiS, Ti_3S_4, and TiO_xS_y (represented as P) phases

category of crystals are, therefore, expected to be relatively poor and those are what have been actually observed. Thus, it is evident that PEC activity of the third category of crystals (Fig. 4) is much improved than either the first or second category. The photoactivity of the third category of crystals is much higher than that of either TiS_2 or TiO_2 (Fig. 1).

A sustained illumination (up to 10 h) of the photoelectrodes, made out of the three categories of crystals, revealed that they do not undergo any visible photocorrosion. In regard to the stability, they were found to compete well with TiO_2 photoanode in all the *p*H values ranging from 1 to 14.

In the present case, since titanium disulphide crystals have been oxidized to produce Ti–S–O phases through selective oxidation at specific temperatures, it is expected that sulphur atoms will be easily substituted by oxygen atoms, due to the higher affinity of titanium towards oxygen. In other words, the loss of sulphur atoms will be compensated by the incorporation of oxygen atoms, producing a Ti–O–S complex. In view of this it is expected that the valence band of TiS_2 will be shifted downwards producing an optimum band gap material, corresponding to an improved response shown in Figure 4. The optimum band gap will presumably correspond to a particular phase with specific values for x and y in TiO_xS_y complex phase. In view of this, the TiO_xS_y phase may be taken as a new modified phase having lower band gap and significant photoresponse in the visible region of solar spectrum. In contrast, $TiO_2(E_g \sim 3.01$ eV) has considerable response only in UV region and exhibits poor response in visible region. The present material typified by TiO_xS_y phase would have a lower band gap and hence better solar spectral response. This would lead to the creation of a higher density of electron-hole pairs and hence a higher photocurrent. Out of the three categories of crystals obtained from TiS_2 after oxidation/reduction treatment, the optimum band gap and hence solar spectral response causing high photocurrent corresponds to the third category of crystals, which are expected to embody TiO_xS_y phases.

Conclusion

In conclusion, it can therefore be said that the present alloy material (TiS_2/TiO_2) with TiO_xS_y phase studied for the first time, represents a new photoanode material. It exhibits improved PEC response. Because of the downward movement of the valence band on oxygen incorporation in TiS_2, the new TiO_xS_y phase is expected to have a lower band gap than TiO_2 and hence a better spectral response. Further studies on the applicability of this new material as photoelectrode in PEC conversion are being carried out presently and the results will be forthcoming.

The authors are grateful to Professor A. R. Verma (NPL, New Delhi), Professor M. V. C. Sastri (University of Madras), Professor G. V. Subba Rao (IIT, Madras), Professor B. Venkatraman (TIFR, Bombay) and Professor M. M. Taqui Khan (CSMCRI, Bhavnagar) for helpful discussions. The present work was financially supported by the Department of Non-conventional Energy Sources (DNES), Ministry of Science & Technology, Government of India, New Delhi.

References

Chitra Pande, Prasad, N., Srivastava, O. N.: Cryst. Res. Technol. **17** (1982) 595
Fujishima, A., Honda, K.: Nature **238** (1972) 37
Gautron, J., Marucco, J. F., Lemasson, P.: Mat. Res. Bull. **16** (1981)

GAUTRON, J., LEMASSON, P., POUMELLEC, B., MARUCCO, J. F.: Solar Energy Mater. **9** (1983) 101
GHOSH, A. K., MARUSKA, H. P.: J. Electrochem. Soc. **124** (1977) 1516
GUYCAMPET, J. V., DOUMERE, J.-P., CLAVERIE, J.: Mat. Res. Bull. **15** (1980) 1135
HARTIG, K. J., GETOFF, N.: Hydrogen Energy Progress VI, New York 1986, p. 546
HARTIG, K. J., GETOFF, N., KOTCHEV, K. D., KANEY, St.: Solar Energy Materials **9** (1983) 167
HOULIHAN, J. F., ARMITAGE, D. B., HOOVLER, T., BONAQUIST, D.: Mat. Res. Bull. **13** (1978) 1205
LEITH, R. M. A., TARSHALL, J. C. J. M.: in: Physics and Chemistry of Materials with Layered Structure, Vol. I, Dordrecht 1977
MASAKAZU ANPO, HIROAKI, N., SUKEYA, K., YUTAKA, K., KAZUNARIDOMEN, ONISHI, T. H.: J. Phys. Chem. **90** (1980) 1633
MURRAY, R. B.: YOFFE, A. D.: J. Phys. C. Solid State Phys. **5** (1972) 3038
PHILLIPS, T. E., MOORJANI, K., MURPHY, J. C., POEHLER, T. O.: J. Electrochem. Soc. **129** (6) (1982) 1210
PRASAD, G., RAO, N. N., SRIVASTAVA, O. N.: in: Progress in Hydrogen Energy, ed. R. P. DAHIYA, Dordrecht 1987
SALVADOR, P.: Mat. Res. Bull. **15** (1980) 1287
SUBBA RAO, S. N., YUN, Y. H., KEREHAW, R., DWIGHT, K., WOLD, A.: Mat. Res. Bull. **13** (1978) 1461
WRIGHTON, M. S.: in: 5th International Conference on Photochemical Conversion and Storage of Solar Energy, Osaka 1984, Japan
ZIELINSKI, S., SOBCZYNSKI, A.: Acta Chim. Hung. **120** (3) (1985) 229

(Received September 7, 1987;
accepted September 12, 1987)

Authors' addresses:

Professor O. N. SRIVASTAVA
Physics Department, Banaras Hindu University
Varanasi-221 005, India

K. S. CHANDRA BABU
Department of Chemistry, Banaras Hindu University
Varanasi-221 005, India

Cryst. Res. Technol. | 23 | 1988 | 4 | 561–566

M. S. Joshi, K. Mohan Prabhu

Department of Physics, Sardar Patel University, Vallabh Vidyanagar, India

Synthesis and Characterization of ZSM-8-type Zeolite Crystals

Procedure for hydrothermal synthesis of ZSM-8-type zeolite crystals, using TEA–OH as a template, in the system having molar composition

$$16.2\,(Na_2O) - Al_2O_3 - 117\,(SiO_2) - 4.8\,(TEA)_2O - 3887\,H_2O$$

at 170 °C is described. Results of characterization of these crystals by XRD, IR, EDAX, SEM, TGA, and adsorption studies are described and discussed. X-ray diffraction studies establish that the crystals are ZSM-8-type with structure similar to that of ZSM-5-type crystals. Crystals have prismatic bipyramidal morphology with average size 18–25 microns. Degased calcined samples adsorb xylenes in the order $p > m > 0$. Pore length per unit cell, estimated from n-hexane adsorption studies, is 46 Å.

1. Introduction

In the recent years a series of zeolites with varying chemical compositions and pore sizes have been synthesized using structure directing role of organic molecules. Structure directing or templating role of organic molecules has been reviewed by Lok et al. Templating is the phenomenon occurring during either the gelation or the nucleation process whereby the organic molecule organizes oxide tetrahedra into a particular geometric topology around itself and this provides the initial building block for a particular structure type. Structure of ZSM-5, ZSM-8, and ZSM-11 serve as best examples for the templating theory, they being formed by the templating role of TPA, TEA, and TBA ions (Lok et al.).

Eversince the discovery of ZSM-8 in the year 1971 (Mobile Oil Corporation), only a few reports are available regarding its synthesis (Sang-Eon et al.; Gabelica et al.). The structure of ZSM-8 zeolite has not yet been reported, and it is suspected to be similar to that of ZSM-5 zeolite (Gabelica et al.; Lok et al.). These crystals give highly promising results in the catalytic conversion of toluene into xylenes (Sang-Eon et al.) The present paper reports the procedure for the synthesis of ZSM-8-type zeolite crystals using TEA. Results of characterization (by different techniques) of the synthesized products are discussed.

2. Experimental procedure

For the preparation of the starting gel sodium silicate or colloidal silica (40% SiO_2 in water), sodium hydroxide, aluminium sulphate, tetraethylammonium hydroxide (20% in water) and double distilled water were used. Gel having molar composition

$$16.2\,(Na_2O) - Al_2O_3 - 117\,(SiO_2) - 4.8\,(TEA)_2O - 3887\,H_2O$$

was prepared using the above reactants, with sodium silicate as the silica source. The pH of the gel was adjusted to 10.5 using dilute sulphuric acid. It was stirred for 3 – 4 hours to ensure homogeneity. Later, it was transferred to a 750 ml teflon lined, glandless induction driven Berghof autoclave. The latter was sealed and heated to 170 °C for 10 days with intermittent stirring. Products obtained at the end of the growth run were separated by decantation, washed several times with double distilled water, filtered out and dried at 100 °C for 12 hours. Some runs were carried out using colloidal silica as the silica source.

3. Characterization

3.1. X-ray powder diffraction studies

X-ray powder diffractograms were recorded in the 2θ range 5 to 70° using CuKα radiation. Figure 1a shows the diffractogram of the calcined sample obtained using colloidal silica, and Figure 1b is the diffractogram for as-synthesized product. From these spectra d values were calculated. Products obtained using sodium silicate as the silica source contain significant amounts of mordenite impurity phase, since the spectrum of this sample contains lines of mordenite (BRECK (a)). Crystals synthesized with colloidal silica as the silica source are similar to those of ZSM-8-type since the calculated d values match well with the reported ones for ZSM-8 zeolites (BRECK (b)).

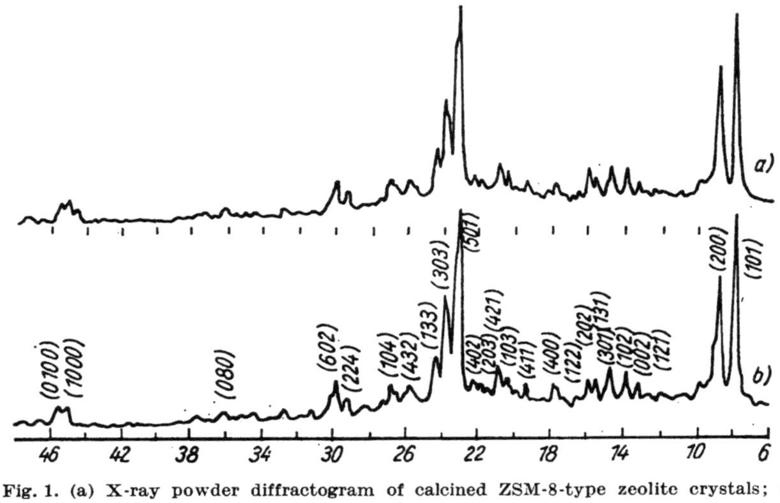

Fig. 1. (a) X-ray powder diffractogram of calcined ZSM-8-type zeolite crystals; (b) X-ray powder diffractogram of as-synthesized ZSM-8-type crystals

Planes were indexed. Table 1 shows the d values, along with the corresponding relative intensities and the Miller indices. These crystals belong to orthorhombic system. Unit cell parameters were estimated as $a = 19.984$, $b = 19.904$ and $c = 13.364$ Å. However, on calcination at 500 °C for 5 h in air, trapped TEA ions get oxidised, resulting in the shrinkage of the initial unit cell parameters to $a = 19.916$, $b = 19.880$, and $c = 13.314$ Å. The peak at 2θ 9.95° which is suspected to be caused by the presence of TEA is found to decrease in height on calcination. The spectrum of the calcined ZSM-8-type sample shows a new diffraction peak at 2θ 44.65°, which is absent in the spectrum of the as-synthesized ZSM-8 type sample. The diffraction pattern of these crystals is similar to that of ZSM-5. A close resemblance is observed

Table 1
X-ray powder diffraction data of ZSM-8 crystals

d/Å	h k l	relative intensity I/I_0
11.163	1 0 1	96
9.992	2 0 0	62
9.025	—	11
7.375	1 2 1	8
6.682	0 0 2	13
6.348	1 0 2	18
5.981	3 0 1	20
5.701	1 3 1	13
5.574	2 0 2	16
5.372	1 2 2	7
4.996	4 0 0	13
4.599	4 1 1	13
4.353	1 0 3	18
4.250	4 2 1	22
4.095	2 0 3	14
4.013	4 0 2	17
3.858	5 0 1	100
3.723	3 0 3	59
3.655	1 3 3	29
3.437	4 3 2	14
3.312	1 0 4	16
3.056	2 2 4	10
2.989	6 0 2	19
2.488	0 8 0	6
2.014	10 0 0	11
1.991	0 10 0	11

in their unit cell parameters. ZSM-5 zeolite also belongs to orthorhombic system, with $a = 20.07$, $b = 19.92$, and $c = 13.42$ Å (as-synthesized). It has been reported by Wu et al.

3.2. Infrared spectral analysis

Infrared spectral analysis, being complimentary to X-ray diffraction, gives a lot of information regarding the building units of the structure. In addition to this, bands arising at higher frequencies (1300–4000 cm^{-1}) give an idea of zeolitic water and absorbed non-structural materials.

IR spectra were recorded using KBr pellet technique in the range 300–4000 cm^{-1}. Figure 2a is the IR spectrum for ZSM-8-type synthesized zeolite, whereas Figure 2b is a spectrum of ZSM-5 zeolite for comparison. Band assignment is made as per Flanigen. The strong bands are caused by the structure insensitive internal tetrahedron vibrations. Here, the band at 1100 cm^{-1} is assigned to the asymmetric T–O stretch, while the medium strong band at 465 cm^{-1} is attributed to the T–O bending mode.

The second class of weak bands are caused by the vibrations in the structure sensitive linkages between tetrahedra. The medium band at 548 cm^{-1} and the very weak band at 600 cm^{-1} are caused by the vibrations of double five-membered rings. The medium weak band at 798 cm^{-1} is related to the symmetric stretch vibrations. The very weak shoulder at 1220 cm^{-1} is assigned to the T–O asymmetric stretch and

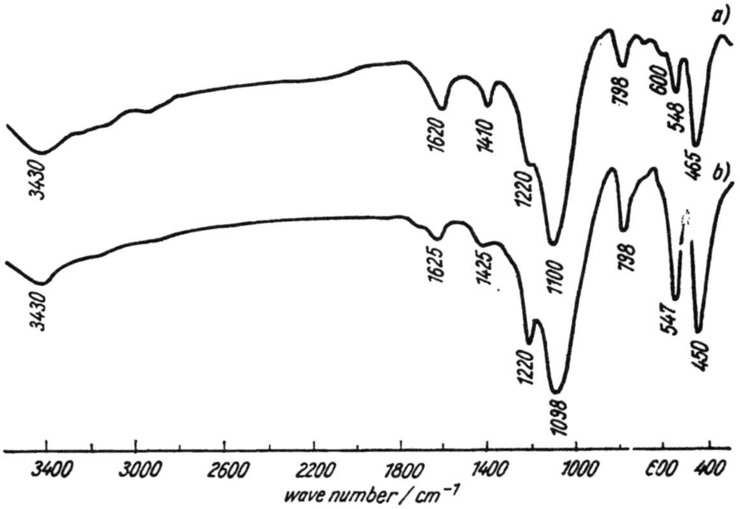

Fig. 2. (a) Infrared spectrum of ZSM-8-type zeolite crystals; (b) Infrared spectrum of ZSM-5-type zeolite crystals

is found only in the zeolite having chains of five-membered rings. Thus, presence of the bands at 548 and 1220 cm^{-1} strongly suggests that these crystals belong to the pentasil zeolite family. The close resemblance of these bands with those of ZSM-5 zeolites proves, beyond doubt, that the structure of ZSM-8-type zeolite must be similar to that of ZSM-5 crystals.

The broad weak band observed at 3430 cm^{-1} is attributed to the stretching vibrations of hydroxyl group of water molecules present in the pores of the zeolites. The weak band at 1620 cm^{-1} results due to the bending vibration of water. The weak band at 1410 cm^{-1} may be effected by the non-structural silica or alumina present in the pores.

3.3. Chemical analysis (EDAX)

From the Energy Dispersive Analysis of X-rays (EDAX) of the ZSM-8-type samples, atomic percents of sodium, aluminium and silicon were determined, and are given below.

Sodium 3.247; aluminium 5.518 and silicon 91.235. The silica to alumina ratio SiO_2/Al_2O_3 is 33.1, which shows that the synthesized ZSM-8-type zeolites are high silica structures ($Si/Al > 10$). Based on 96 TO_2 (T means Si or Al) for the unit cell of ZSM-5 structure, unit cell composition of the ZSM-8-type zeolite is estimated from EDAX results as under

$$(Na_2O)_{1.61}\,(Al_2O_3)_{2.74}\,(SiO_2)_{90.52}\,.$$

3.4. Morphology and particle size

Using the scanning electron microscope, morphology of the synthesized products was studied. ZSM-8-type crystals exhibit prismatic bipyramidal morphology (Fig. 3). Their average size is estimated to be 18—25 microns. The impurity mordenite crystals are needle-shaped.

3.5. Thermal analysis (TGA)

Thermogravimetric analysis gives a quantitative measure of zeolitic water and organic species trapped in the pores during the growth of the crystals. For TGA, crystals were

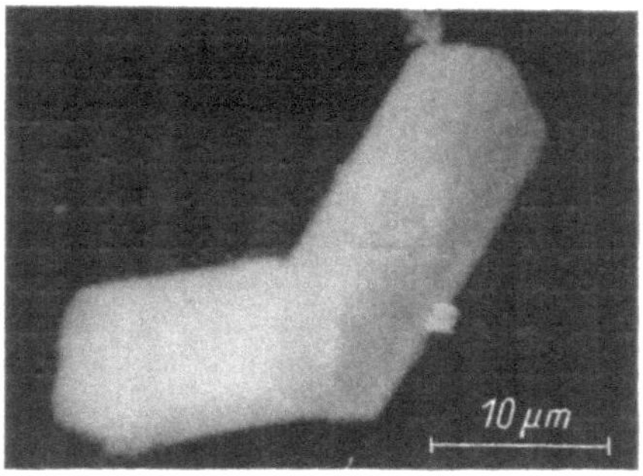

Fig. 3. A scanning electron micrograph showing prismatic bipyramidal ZSM-8-type crystals

heated at the rate of 10 °C/min in air from 25 to 650 °C. Weight loss curves were obtained for pure and mixed phases. Figure 4 shows the weight loss curve for ZSM-8 type zeolite crystals. From 25 to 350 °C weight loss is attributed to the desorption of zeolitic water, which amounts to 4.92 weight percent. Above 350 °C trapped TEA ions get oxidised as evidenced by a desorption step. The weight loss between 350 and 650 °C is 5.68 percent. Hence, the number of water molecules per unit cell is estimated to be 17.8, and TEA 2.81. The single ethane chain of TEA is 4.71 Å Total chain length for 2.81 molecules of TEA gives an idea of total pore length per unit cell. This turns out to be 46.03 Å.

From EDAX and TGA complete unit cell composition is worked out to be

$$(Na_2O)_{1.61}\,(Al_2O_3)_{2.74}\,(SiO_2)_{90.52}\,(TEA)_{2.81}\,(H_2O)_{17.77}\,.$$

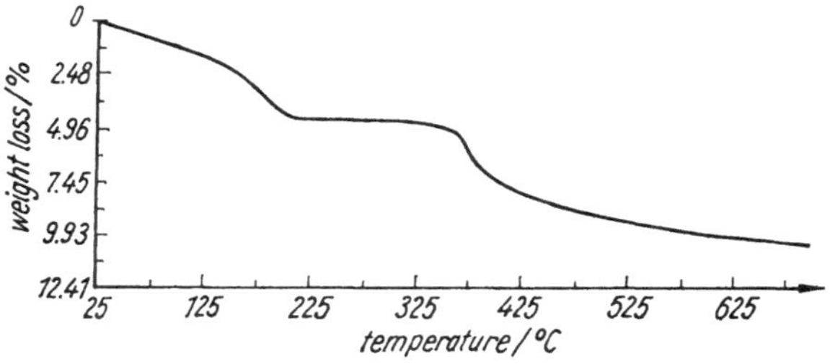

Fig. 4. Thermogravimetric weight loss curve of ZSM-8-type zeolite crystals

3.6. Adsorption studies

Using McBain quartz spring balance, adsorption capacities of calcined ZSM-8 type zeolites for xylenes, n-hexane and water were estimated. Calcined samples were degased at 10^{-3} torr, and adsorption studies were carried out. Crystals exhibit selective adsorption for p-xylene over m-xylene and o-xylene. Total adsorption capacities (weight percents) for different molecules are given below, at 24 °C, at their respective saturation vapour pressures.

Molecules	Adsorption capacity (wt%)
p-xylene	8.75
m-xylene	5.80
o-xylene	4.80
n-hexane	6.87
water	7.42

Total number of n-hexane molecules adsorbed per unit cell is estimated to be 4.64. The length of one n-hexane molecule is 9.92 Å. Hence, length of the pore per unit cell is 46.03 Å, which is in good agreement with the reported one by GABELICA et al.

It is concluded that ZSM-8-type zeolites synthesized in the present investigation are almost similar to ZSM-5 zeolites in several respects. Crystals of ZSM-8-type can be easily synthesized using colloidal silica as the silica source, whereas the use of sodium silicate leads to incorporation of mordenite impurity phase in ZSM-8. These synthesized ZSM-8-type crystals exhibit prismatic bipyramidal morphology. Their average size is estimated to be 18—25 microns. n-hexane adsorption shows that total pore length is 46.03 Å per unit cell.

The authors are grateful to the Council of Scientific and Industrial Research, New Delhi, India for the financial assistance in the form of a research project for carrying out this work.

References

BRECK, D. W. (a), (b): Zeolite Molecular Sieves, New York 1974, pp. 363, 373
FLANIGEN, E. M.: Zeolite Chemistry and Catalysis (Ed. J. A. RABO) Adv. Chem. Ser. 1976, Ch. 2, p. 171
GABELICA, Z., DEROUANE, E. G., BLOM, E.: Applied Catalysis **5** (1983) 109
LOK, B. M., CANNAN, R. R., MESSINA, C. A.: Zeolites **3** (1983) 282
Mobil Oil Corp.: Neth. Pat. 70 14807 (1971)
SANG-EON, P., HAKZE, C.: Proc.-Pac. Chem. Eng. Cong., 3rd **2** (1983) p. 197
SANG-EON, P., HAKZE, C.: Chem. Eng. Commun. **34** (1—6) (1985) 137
WU, E. L., LAWTON, S. L., OLSON, D. H., ROHRMANN, R. C. (Jr.), KOKOTAILO, G. T.: J. Phys. Chem. **83** (1979) 2777

(Received September 15, 1987;
accepted October 6, 1987)

Authors' address:

Prof. M. S. JOSHI, K. MOHAN PRABHU
Department of Physics, Sardar Patel University
Vallabh Vidyanagar-388120 Gujarat, India

Cryst. Res. Technol.	23	1988	4	567–575

P. Deus, M. John, H. Krause, H. Sauermann

Sektion Physik, Sektion Metallurgie und Werkstofftechnik und
Sektion Maschinen- und Energietechnik der Bergakademie Freiberg

Calculation of Temperature Fields in a Cylindrical Heater System

A model for calculation of time dependent temperature fields in a cylindrical furnace used for crystal growth experiments is described. The comparison of the calculated and the experimental temperature profiles shows good agreement in the stationary case and the suitability of the model for the prediction of static temperature fields on the basis of given heating power distributions, and vice versa.

Ein Modell zur Berechnung von zeitabhängigen Temperaturfeldern in bei der Kristallzüchtung eingesetzten Rohröfen wird beschrieben. Der Vergleich von berechneten und experimentellen Temperaturprofilen zeigt im stationären Fall gute Übereinstimmung und die Eignung des Modells zur Vorhersage statischer Temperaturfelder bei gegebenen Heizleistungsverteilungen und umgekehrt.

1. Introduction

Heater systems with rotational symmetry are used widely as furnaces in science and technology. Modern developments, e.g. in the field of crystal growth or semiconductor devices demand increased stability of complicated temperature fields and, furthermore, a controlled dependence on time.

From the point of view of control engineering, such a furnace is a multivariable control system the behaviour of which is considerably influenced on by the geometry of the furnace and the materials used for construction. Therefore, geometry and materials used in furnace construction should be chosen carefully with respect to the static and dynamic thermal conditions desired for a special purpose.

It is the aim of this paper to report on a model for calculation of time dependent temperature fields in a cylindric furnace used for crystal growth experiments with the objective to predict its static and dynamic behaviour and, thus, to optimate development and manufacture of furnace and control equipment. Then, the static behaviour of the model is shown by comparison with experimental results.

2. Experimental

For comparison of numerical results of the model and experimental reality a simple multiple zone furnace (Fig. 1) was used. The heater elements (1) of each zone consist of Kanthal A1 and were surrounded by concentric refractory elements (2) for stability. The external thermal insulation body (3) is composed of several layers of ceramic fibre paper and wrapped up by an aluminium mantle (4). The axial boundaries of furnace are disks (5) of refractory material with holes for Pt–PtRh thermocouples (6) for temperature measurement. Each heater element was supplied with electric power of maximally 0.2 kW suffi-

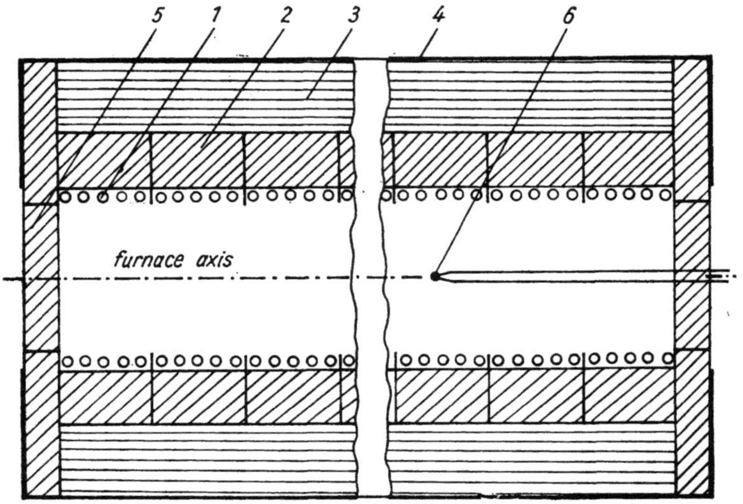

Fig. 1. Multiple zone furnace used for experimental investigations (see text)

cient for an internal furnace temperature ϑ up to 1300 °C. The error of temperature measurement was ± 2 K, the heating power of the zones was held constant within $\pm 1‰$.

3. General description of model

The control scheme of the furnace is shown in Figure 2. Each zone (number k) is controlled by a TIRC unit, from which the control quantity Y_k (heat flow into the zone due to electrical heating or cooling) is given. Y_k regulates the controlled variable X_k (temperature) which again is fed back into the TIRC unit. The field of external temperature acts in the form of a boundary condition as interference function. Due to internal heat currents in the furnace (thermal conduction, radiation, and convection) each control quantity Y_k influences all controlled variables X_j. Thus, the furnace represent a multivariable control system (Fig. 3) with the transfer functions G_{jk}.

From the physical point of view the coupling of zones is realized by the heat flows mentioned. We have to solve within the furnace body an inhomogeneous heat conduction problem represented by equ. (1).

$$c\varrho \frac{d\vartheta}{dt} - \operatorname{div}(\lambda \operatorname{grad} \vartheta) = \dot{q} \,. \tag{1}$$

The solution desired is the space and time dependence of temperature $\vartheta(\boldsymbol{r}, t)$ in result of the heat source density $\dot{q}(\boldsymbol{r}, t)$ consisting of the interior heat source or of the

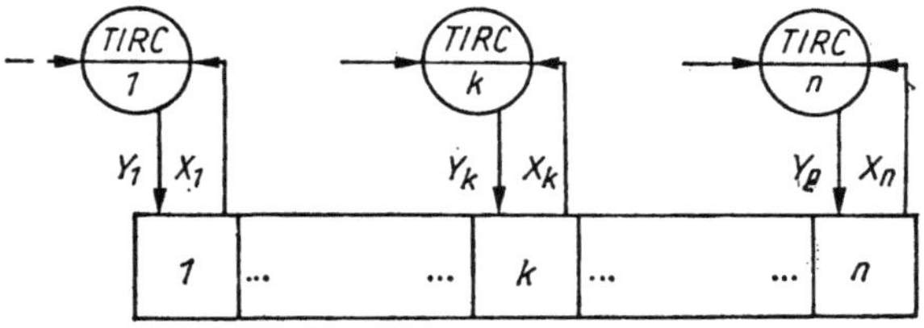

Fig. 2. Furnace control scheme, TIRC-temperature indicating, recording, and controlling equipment

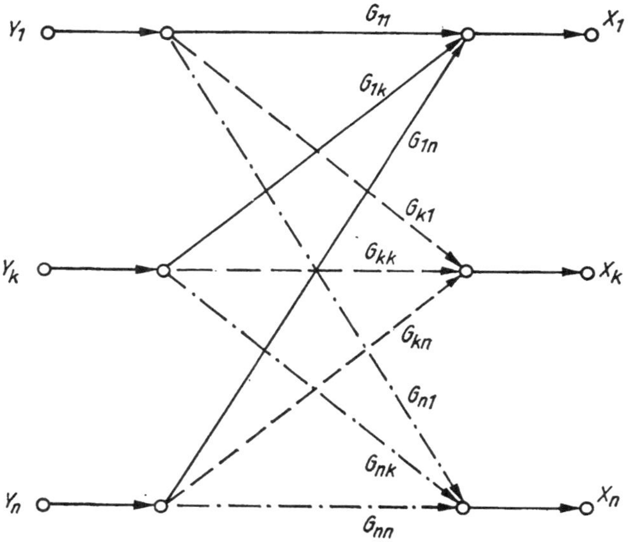

Fig. 3. Signal flow diagram of the furnace

drain density due to heating or cooling, respectively. At the boundaries, heat flows due to radiation and convection have to be taken into consideration. In equ. (1) $c(\mathbf{r}, \vartheta)$ is the specific heat, $\varrho(\mathbf{r}, \vartheta)$ the mass density and $\lambda(\mathbf{r}, \vartheta)$ the heat conductivity.

The solution of equ. (1) has been found by means of the balance method (REISSMANN; LINK; KRAUSE). The discretisation of space is realized in the case of cylindrical symmetry by cutting the furnace body into several elements in form of hollow cylinders with thicknesses Δr_1 and lengths Δz_1 according to Figure 4. According to Figure 5 the time discretisation of element 1 (temperature ϑ_1) is possible using the

forward $(\vartheta_{l,n+1} - \vartheta_{l,n})/\Delta t$,

backward $(\vartheta_{l,n} - \vartheta_{l,n-1})/\Delta t$,

or central $(\vartheta_{l,n+1} - \vartheta_{l,n-1})/2\,\Delta t$

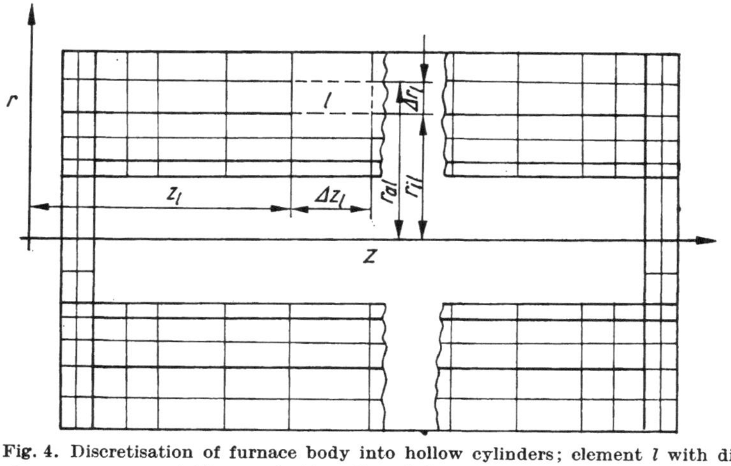

Fig. 4. Discretisation of furnace body into hollow cylinders; element l with dimensions Δr_l, Δz_l; r_{al} resp. r_{il}: outside resp. inside radius of element l

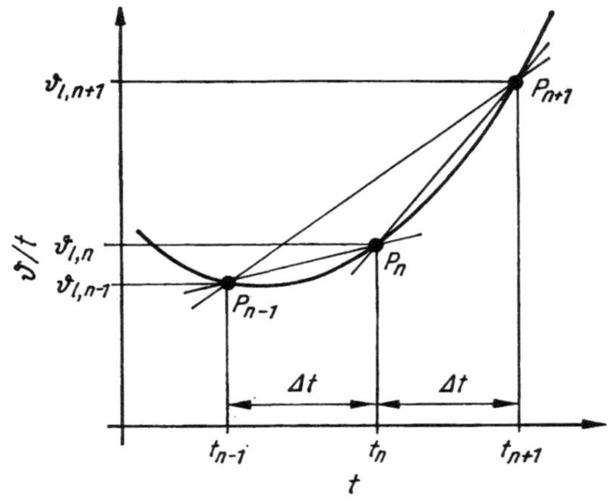

Fig. 5. Possibilities for time discretisation: $P_{n+1} - P_n$ forward, $P_n - P_{n-1}$ backward, and $P_{n+1} - P_{n-1}$ central difference quotient

difference quotient, respectively (SMITH), all of which have certain advantages and disadvantages with respect to stability, precision, and numerical expense of the solution procedure. In this work, a modified central quotient according to CRANK-NICOLSON has been used.

4. Heat balances of furnace elements

Each furnace element is represented by an internal nodal point l with the coordinates $z_{Nl} = z_l + \Delta z_l/2$ and r_{Nl} (Fig. 6). The temperature ϑ_l of element l is the temperature which the whole element would have after a complete heat exchange within itself. Due to the radial temperature dependence ϑ_l is present in reality at the nodal radius

$$r_{Nl} = r_{il}\, e^{-1/2} \left(\frac{r_{al}}{r_{il}}\right)^{r_{al}^2/r_{al}^2 - r_{il}^2} . \tag{2}$$

In in the case of $r_{al}/r_{il} \approx 1$, r_{Nl} is not far from the arithmetic mean value $r_{Ml} = (r_{al} + r_{il})/2$.

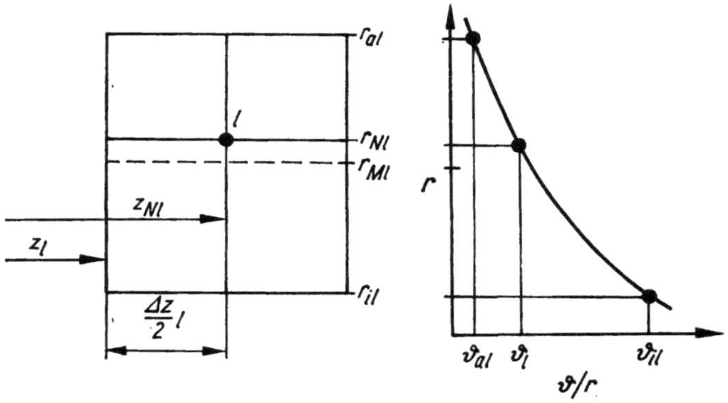

Fig. 6. Internal nodal point of element l and radial temperature dependence within this element; r_{Nl} nodal, r_{Ml} mean radius, resp.

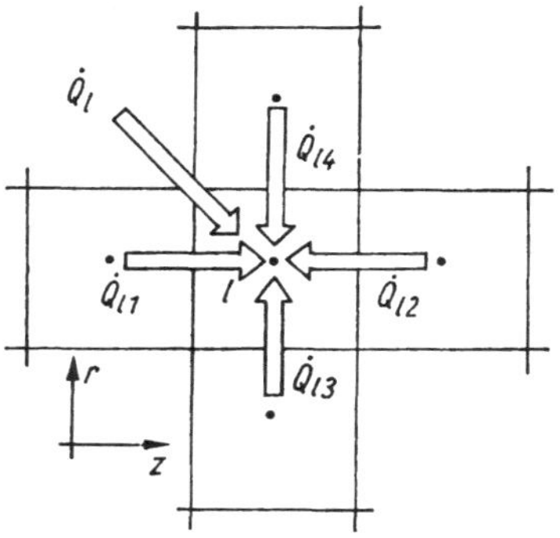

Fig. 7. Heat balance of element l (see text)

The heat current balance for an internal nodal point 1 (see Fig. 7) is given according to the energy conservation by equ. (3)

$$\frac{dH_1}{dt} = \sum_{j=1}^{4} \dot{Q}_{lj} + \dot{Q}_{lj} \, . \tag{3}$$

H_l is the enthalpy of element l, $\dot{Q}_{lj}$ the heat current from the neighbouring element j into element l and $\dot{Q}_l$ the heat current into element l due to external heating or cooling. Proceeding to the difference formalism equ. (3) yields

$$\Delta H_l = (\sum_{j=1}^{4} \dot{Q}_{lj} + \dot{Q}_l) \, \Delta t \, . \tag{4}$$

Using for simplicity the forward difference quotient here, the variables in equ. (4) can be expressed as follows:

$$\Delta H_l = m_l c_l (\vartheta_{l,n+1} - \vartheta_{l,n}) \, , \tag{5}$$

where m_l and c_l are mass and specific heat of element l, respectively

$$\dot{Q}_{lj} = \Lambda_{lj} (\vartheta_{j,n} - \vartheta_{l,n}) \, , \tag{6}$$

where Λ_{lj} is the reciprocal heat resistance between the elements l and j.

The Λ_{lj} have to be distinguished in accordance with the direction of the heat current. For axial heat exchange ($j = 1, 2$)

$$\Lambda_{lj} = \frac{2\pi (r_{al}^2 - r_{il}^2)}{\dfrac{\Delta z_j}{\lambda_j} - \dfrac{\Delta z_l}{\lambda_l}} \tag{7}$$

is valid with λ_j and λ_l being the heat conductivity of element j and l, respectively. $j = 3$ (heat current from the inside neighbour of element l) yields

$$\Lambda_{lj} = \frac{2\pi \, \Delta z_l}{\dfrac{1}{\lambda_l} \ln \dfrac{r_{Nl}}{r_{il}} + \dfrac{1}{\lambda_j} \ln \dfrac{r_{il}}{r_{Nj}}} \, , \tag{8}$$

and $j = 4$ (outside neighbour of l) gives

$$\Lambda_{lj} = \frac{2\pi \, \Delta z_l}{\dfrac{1}{\lambda_j} \ln \dfrac{r_{Nj}}{r_{al}} + \dfrac{1}{\lambda_1} \ln \dfrac{r_{al}}{r_{Nl}}} \, . \tag{9}$$

For boundary elements the expressions (7)—(9) have to be changed, omitting in the denominator the corresponding summands with the index j. These elements possess

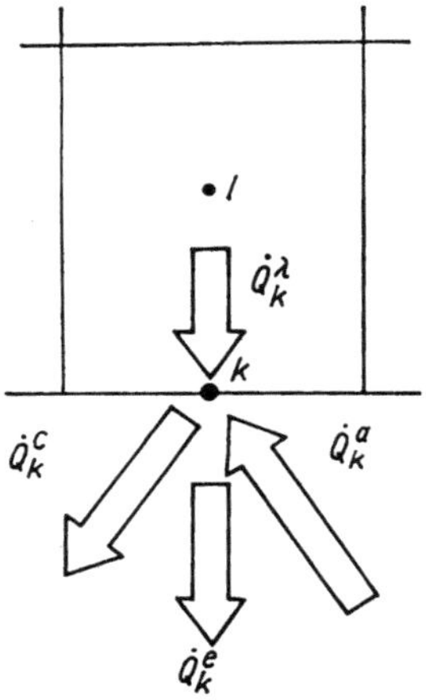

Fig. 8. Heat balance of boundary node k (see text)

an additional node (number k) at the boundary. For these boundary nodes, the heat currents are balanced, i.e. the heat current $\dot{Q}_{lk} = \dot{Q}_k^\lambda$ due to conduction from the internal node has the same value as the sum of the external heat currents due to convection $\dot{Q}_k^c$ and radiation $\dot{Q}_k^r = \dot{Q}_k^a - \dot{Q}_k^e$ (absorbed respectively emitted heat flow, see Fig. 8)

$$\dot{Q}_k^\lambda - \dot{Q}_k^c + \dot{Q}_k^a - \dot{Q}_k^e = 0 . \tag{10}$$

The reflected part of the incident radiation has not to be taken into account in the energy balance of the boundary nodes.

$\dot{Q}_k^\lambda$ has to be calculated according to the modified equations (6)—(9). $\dot{Q}_k^c$ can be expressed as

$$\dot{Q}_k^c = A_k(a_{1k} + a_{2k}\vartheta_k)(\vartheta_u - \vartheta_k) . \tag{11}$$

A_k is the outside area of the element containing node k, a_{1k} and a_{2k} are quantities describing the heat transfer trough A_k due to convection, and ϑ_u the temperature of the environment. In the case of high temperatures the heat flow due to convection is (at interior boundary nodes) considerably smaller than that due to radiation and will be neglected here.

The Stefan-Boltzmann law yields for the interior boundary nodes

$$\dot{Q}_k^e = A_k \varepsilon_k T_k^4 \tag{12}$$

with ε_k being the emissivity an T_k the absolute temperature of node k. For calculation of $\dot{Q}_k^a$ the interaction between the different nodes has to be taken into consideration using the so-called resolvent method (see ISACHENKO et al.).

The direct heat flow $\dot{Q}_{ik}^d$ by radiation from an interior boundary node i to the node k under consideration can be written as

$$\dot{Q}_{ik}^d = \varphi_{ik}\dot{Q}_i^e \tag{13}$$

with the direct irradiation factor φ_{ik} (sight factor) being unambiguously determined by the geometry of the furnace. On the other hand, radiation from node i reaches node k indirectly via one or more reflections at all remaining nodes as well. Therefore, the total irradiation factor Φ_{ik} describes the total part $\dot{Q}_{ik}^t$ of radiation emitted by node i reaching node k.

$$\dot{Q}_{ik}^t = \Phi_{ik}\dot{Q}_i^e . \tag{14}$$

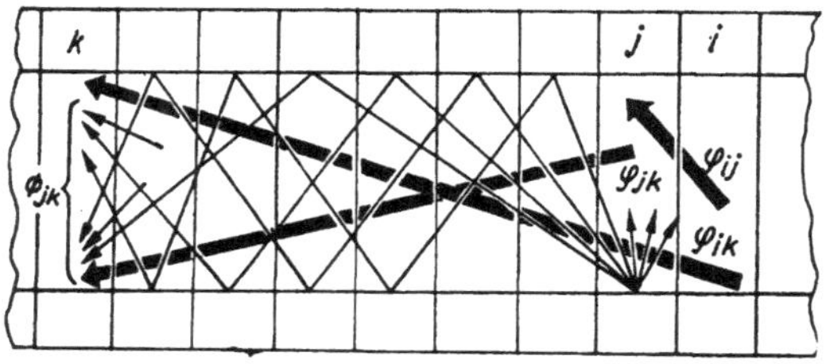

Fig. 9. Direct irradiation factors φ_{ik}, φ_{jk}, and φ_{ij}, and total irradiation factor Φ_{jk} including φ_{jk} and contributions due to multiple reflections

The unknown total irradiation factors Φ_{ik} can be calculated according to Figure 9 as follows: Node j receives from node i the direct flow $\dot{Q}^d_{ij} = \varphi_{ij}\dot{Q}^e_i$, the part $r_j\dot{Q}^d_{ij}$ of which is reflected again. The part $\Phi_{jk}r_j\dot{Q}^d_{ij}$ of the reflected flow reaches node k. The heat flow $\dot{Q}^t_{ik}$ (equ. (14)) is composed by the direct flow (equ. (13)) and the reflected flows from all remaining elements j. From this follows the system of equations (15) by means of which all Φ_{ik} can be calculated.

$$\Phi_{ik} = \varphi_{ik} + \sum_{j=1}^{N} \Phi_{jk}r_j\varphi_{ij} \,. \tag{15}$$

Now, $\dot{Q}^a_k$ follows as

$$\dot{Q}^a_k = \alpha_k \sum_{i=1}^{N} \Phi_{ik}\dot{Q}^e_i \,, \tag{16}$$

if α_k is the absorption factor of node k. Besides the temperature dependence of convection (equ. (11)), the temperature dependence of heat conductivity (BOLDT et al.) and specific heat capacity are used as follows

$$\lambda_1(T) = \frac{a_{31}}{T} + a_{41}\sqrt{T} + a_{51}T^3 \,, \tag{17}$$

$$c_1(\vartheta) = a_{61} + a_{71}\vartheta + a_{81}\vartheta^2 \,. \tag{18}$$

The quantities $a_{11}, \dots, a_{81}$ of equations (11), (17) and (18) are calculated by means of data taken from different tables or information given by the producers.

Now, from the heat flow balances a nonlinear system of equations can be drawn up, the solution of which yields after an iterative linearisation the nodal temperatures depending on time. For comparison with the experimental equipment shown in Figure 1 the temperatures of the interior boundary nodes have been used.

5. Results

The calculations were carried out by means of a computer BC 5130. Figure 10 shows the difference of actual and true result (infinite number of iterations) depending on the iteration limit $\Delta\vartheta_0$ given by the operator. Finally, an iteration limit of 0.01 K was chosen corresponding to an error in temperature calculation of ± 0.8 K. In Figure 11 the theoretical and experimental temperature characteristics in the centre of the furnace are compared, if the furnace is heated uniformly. The best agreement of the model with experimental values is given near 1120 °C. The small difference of both curves should be explained by the unreliability of the material data and their temperature dependence in this temperature region. A better insight into the sensivity of the model towards structural modifications is given by comparison of experimental and theoretical temperature profiles. In the example given in Figure 12 the power distribution was chosen for the establishment of a small temperature minimum

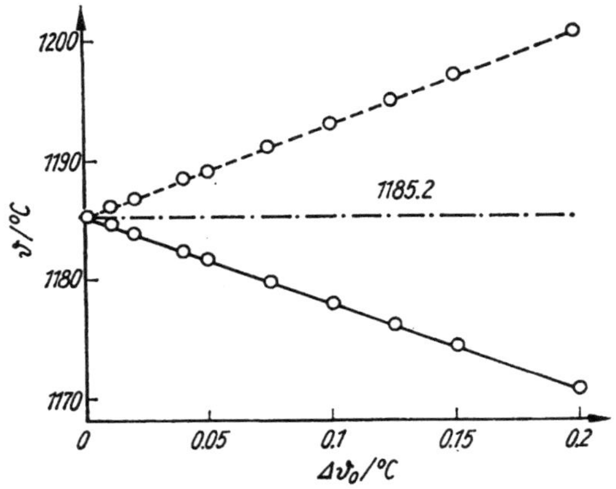

Fig. 10. Dependence of the result of calculation on iteration limit $\Delta\vartheta_0$ for lower (solid line) and higher (dashed) starting values of temperature

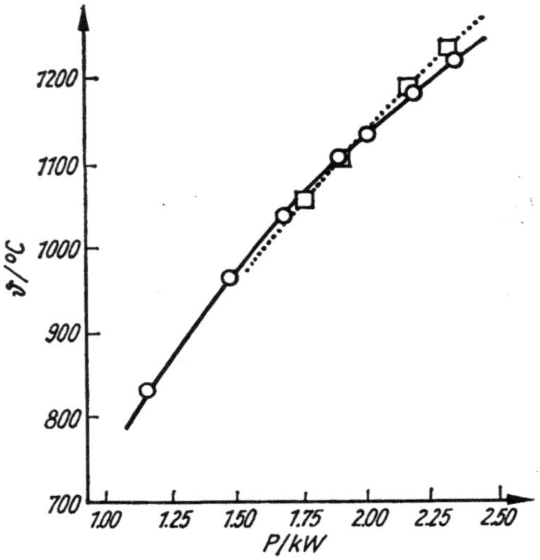

Fig. 11. Temperature characteristic of the centre of the uniformly heated furnace, solid line – experimental, dotted – calculated

in the furnace center. There is an excellent structural agreement between model and experiment in the main part of the furnace. The reasons for the higher theoretical temperatures in the outer zones, the influence of which we are not able to estimate quantitatively with sufficient precision so far, are the following: parasitic heat flows through axial furnace walls, lacking knowledge of material data, errors in temperature measurement due to heat loss at the thermocouple by radiation and conduction (STEINHARDT). Figure 13 shows the establishment of a steep temperature gradient by switching off five heating zones. Again with exception of the outer furnace zones, the model is in good agreement with the experimental values in this case as well. As has been shown by these results, the model described gives a good representation of the experimental conditions in the static case and can be used for prediction of tem-

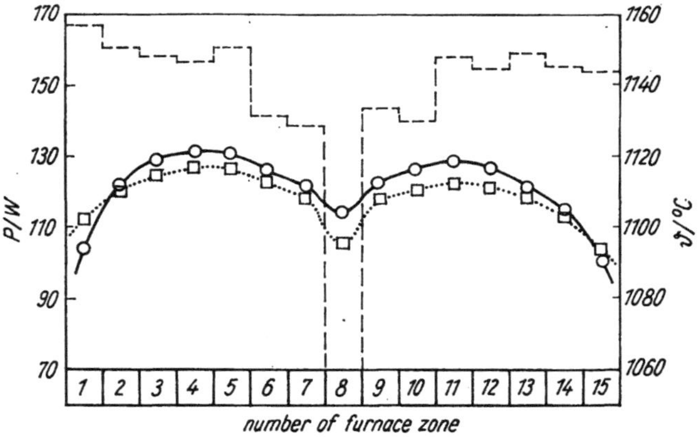

Fig. 12. Experimental (solid line) and calculated (dotted) temperature profile of the furnace for a power distribution (dashed) chosen in order to establish a minimum in the furnace centre

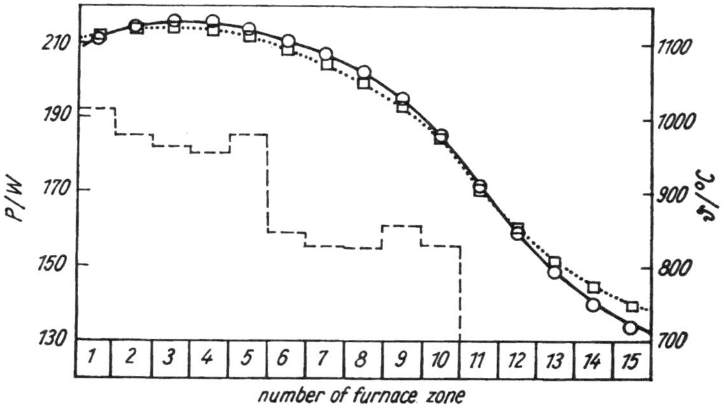

Fig. 13. Temperature profile produced by switching off heating zones 11 – 15, symbols see Figure 12

perature profiles from given power distributions, and vice versa. Furthermore, it can be employed for optimisation of furnace constructions. On the dynamic behaviour of the model will be reported elsewhere.

The authors have to thank Prof. Dr. K. Hein, Prof. Dr. P. Metzing, Prof. Dr. H. A. Schneider, Prof. Dr. R. Steinhardt, and Dr. sc. techn. G. Walter for valuable discussions and S. Rabe and P. Hollstein for technical assistance.

References

Boldt, R., Guschmann, A., Schlegel, E.: Silikattechnik **36** (1985) 119

Crank, J., Nicolson, P.: Proc. Cambridge Phil. Soc. **43** (1947) 50

Isachenko, V. P., Osipowa, V. A., Sukomel, A. S.: Teploperedacha, Moscow 1975

Krause, H.: Diploma work, Bergakademie Freiberg, Sektion Maschinen- und Energietechnik 1986

Link, R.: Dissertation A, Technical University Dresden 1982

Reissmann, C.: Dissertation B, University Rostock 1979

Smith, G. D.: Numerical Solution of Partial Differential Equations, London 1965

Steinhardt, R.: private communication

(Received September 21, 1987; accepted October 2, 1987)

Authors' addresses:

Dr. P. Deus
Sektion Physik, Bergakademie Freiberg
Cottastr. 4
Freiberg
DDR-9200

Dipl.-Ing. H. Krause, Dipl.-Ing. H. Sauermann
Sektion Maschinen- und Energietechnik, Bergakademie Freiberg
Zeunerstr. 7
DDR-9200

Dr. M. John
Sektion Metallurgie und Werkstofftechnik, Bergakademie Freiberg
Leipziger Str. 24
Freiberg
DDR-9200

Cryst. Res. Technol.	23	1988	4	577–581

G. H. Nalbandyan, V. A. Nemenov

Condensed Media Physics Institute of Yerevan State University, USSR, and NPO "Monokristallreaktiv", Kharkov, USSR

Transient Regimes in the Growth of Large-diameter Crystals

The behaviour of crystallization rate and interface position under the change of Czochralski growth system parameters is theoretically and experimentally investigated for the case of large diameter crystals growth. It is shown that the large value of crystal diameter leads to the inertia and nonmonotonic character of system relaxation after the perturbation. The possibility of hesitative nonstability is discussed. The transient regimes in the growth of 30 cm diameter KCl single crystals are investigated experimentally. The behaviour of crystallization rate differs from the case of small diameter crystals and corresponds to the results of theoretical analysis.

Теоретически и экспериментально исследовано поведение скорости кристаллизации и положения фронта кристаллизации при возмущении параметров ростовой системы Чохральского в случае выращивания кристаллов большого диаметра. Показано, что большое значение диаметра кристалла приводит к инерционности и немонотонности релаксации системы. Найдены условия возникновения колебательной неустойчивости процесса. Экспериментальные исследования проводились при выращивании монокристаллов KCl диаметром более 30 см. Поведение скорости кристаллизации отлично от случая роста кристаллов малого диаметра и соответствует результатам теоретического анализа.

The growth of perfect single crystals doesn't only require a careful choice of the growth regime, but a thorough maintaining of selected regime as well. An accidental change of growth system parameters and inadequate operation lead to fluctuations of crystallization rate. The crystallization rate v is one of the most important growth parameters, its fluctuations are the basic reason of inhomogeneities emergence in the growing crystal. That's why the effect of Czochralski system parameters change on the crystallization rate is discussed in the present paper.

The transient regimes during the growth are determined by the relaxation processes, connected with the shift of the interface from the equilibrium position. The most intense relaxation realizes due to the change of temperature gradient in liquid phase because of the change of thermal boundary layer thickness h. The liquid phase relaxation time t_L equals (Hangleiter, Spaeth; Leybovich, Pogodin)

$$t_L = \frac{Lh_0}{\lambda_L G_L} = \frac{\delta H}{\delta v}. \tag{1}$$

Here δH is the change of interface position, δv is the change of the crystallization rate, L is the latent heat of crystallization, G_L is the stationary value of temperature gradient in liquid phase at the interface, λ_L is the melt thermoconductivity.

The relaxation (1) affects on the crystallization rate with the delay

$$\tau_L = \frac{h_0^2}{\varkappa_L}, \tag{2}$$

where $\varkappa_L$ is the melt temperature conductivity.

The relaxation due to the change of temperature gradient in crystal has a characteristic time

$$t_S = \frac{Ll_0}{\lambda_s G_s} \tag{3}$$

and delay

$$\tau_s = \frac{l_0^2}{\varkappa_s}. \tag{4}$$

Here l_0 is the stationary value of the thermal flow lines length l from the interface to the crystal surface, λ_s, $\varkappa_s$ and G_s are the thermoconductivity, temperature conductivity and temperature gradient, corresponding to crystal. The system relaxation has a characteristic time

$$\tau_r = \left(\frac{1}{t_L} + \frac{1}{t_s}\right)^{-1}.$$

If the diameter of growing crystal is small enough, the lower boundary of thermal layer is completely or partially fixed, the change δH of interface position leads to the (equal or less) change δh of layer thickness, the crystallization rate changes and the relaxation through the liquid phase realizes. In Czochralski systems $h_0 \ll l_0$ and if the liquid phase relaxation occurs, the solid phase relaxation can be neglected. If the diameter of growing crystal isn't small, forced convection is intensive and the interface is sunk into the melt, the change of interface position doesn't lead to the change of thermal layer thickness. In this case the liquid phase relaxation doesn't realize and the system relaxation occurs through the solid phase. If the crystal diameter is large enough, the relaxation delay τ_s can be comparable with the relaxation time τ_r. The high value of relaxation delay τ_s leads to some specific features of system behaviour at transient regimes.

The transient regimes under the stepwise changes of system parameters were experimentally investigated during the growth of KCl single crystals of 30 cm diameter in Czochralski growth system, analogous to GORILETSKY et al. The axial symmetry of thermal fields was ensured by simultaneous rotation of crystal and crucible. The Reynolds number was more than 10^3. The melt level in the crucible was maintained over the whole growth process by continuous feeding. The central flat part of the interface was sunk into 30—50 mm depth. The parameters of growth system (pulling rate or heater power) were stepwise perturbated when the crystal body grows. The crystallization rate behaviour is determined a posteriori from the light scattering centres distribution along the grown crystal. The correlation between the crystallization rate behaviour and scattering centres density is verified at some special experiments. It is found out that the centres distribution follows the crystallization rate behaviour with the delay less than 10 min.

Our investigations give the large values of relaxation time τ_r and delay of relaxation τ_s in our system. The crystallization rate behaviour significantly differs from the case of small diameter crystals growth (NALBANDYAN 1982).

The behaviour of the crystallization rate, found by the one-dimensional heat transfer approximation with a perturbation technique, has the convolution form (NALBANDYAN 1984)

$$\delta v(t) = \delta v_p(t) * F_v(t) + \delta T(t) * F_T(t) + \delta H(t) * F_H(t) .$$

Here $\delta v_p(t)$ and $\delta T(t)$ are the perturbations of the pulling rate v_p and melt bulk temperature T respectively, $F_v(t)$, $F_T(t)$ and $F_H(t)$ are the spectral sensitivities.

Let's consider the behaviour of crystallization rate under the stepwise change of the pulling rate

$$\delta v(t) = \delta v_p \cdot \int_0^t F_v(t)\, dt .$$

Here δv_p is the amplitude of the step. For the initial time interval, when $t \ll \tau_s$, we have

$$\delta v(t) = \delta v_p \frac{\sqrt{\tau_s}}{\tau_r} \left[\sqrt{t} \exp\left(-\frac{\tau_s}{4t}\right) - \sqrt{\tau_s} \operatorname{Erfc}\left(\sqrt{\frac{\tau_s}{4t}}\right)\right]. \tag{5}$$

When $I_s \ll t \ll \tau_1$ we have

$$\delta v(t) = \delta v_p \left[1 - \frac{(\tau_1 + \tau_r) \exp\left(-\frac{t}{\tau_r}\right) - 2\tau_r \exp\left(-\frac{t}{\tau_1}\right)}{\tau_1 - \tau_r}\right]. \tag{6}$$

Here

$$\tau_1 = \frac{4\varkappa_s}{v_0^2} = \frac{4\tau_r^2}{\alpha\tau_s}, \qquad \alpha = \frac{v_0}{v_{max}} .$$

v_0 is the average growth rate, $v_{max} = \frac{\lambda_s G_s}{L}$ is the maximaum possible average growth rate.

If the relaxation delay τ_s is comparable with the relaxation time τ_r, the behaviour of crystallization rate is nonmonotonic. Under the change of the pulling rate the difference between the crystallization and pulling rates leads to the change of interface position. When the interface reaches its new stationary position, the crystallization rate is — because of delay τ_s — still less than the pulling rate and the interface keeps on changing its position until v equals v_p. At that moment the interface turned to be shifted from stationary position, so the interface has to return. The behaviour of crystallization rate and interface position under the change of pulling rate is represented in Figure 1. The case of inertialess relaxation is represented with the dotted lines.

The steady state regime realizes if $\tau_s < \tau_1$. If $\tau_s = \tau_1$ the crystallization rate and interface position hesitate without damping. The period of hesitations is of $\sqrt{2\tau_s\tau_r}$ order. If $\tau_s > \tau_1$ the system becomes unstable. The hesitative nonstability realizes for very large diameters of the growing crystal and large values of crystallization rate.

The stepwise change δH of crystal position gives the change of crystallization rate

$$\delta v(t) = \delta H \cdot F_v(t) .$$

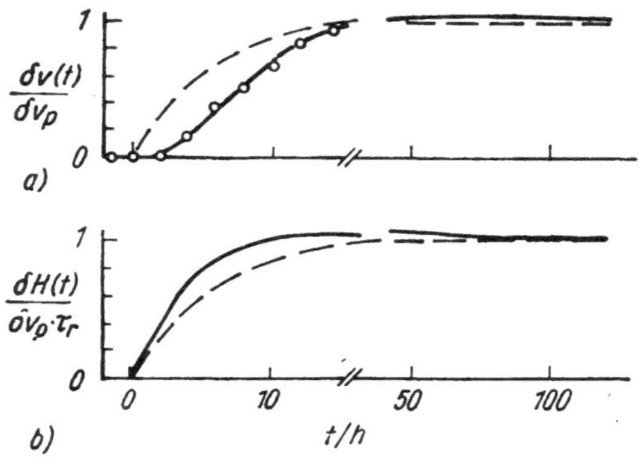

Figs. 1 – 3. The behaviour of crystallization rate (a) and interface position (b) under the stepwise change of pulling rate (Fig. 1), crystal position (Fig. 2) and melt temperature (Fig. 3). The dotted lines correspond to the inertialess relaxation. The experimental results are represented in Figs 1a and 3a with the circles

At $t \ll \tau_s$

$$\delta v(t) = \frac{\delta H}{\tau_r} \sqrt{\frac{\tau_s}{4t}} \exp\left(-\frac{\tau_s}{4t}\right).$$

At the time interval $\tau_s \ll t \ll \tau_r$ the behaviour of crystallization rate is nonmonotonic

$$\delta v(t) = \frac{\delta H}{\tau_1 - \tau_r} \left[\frac{\tau_s + \tau_r}{\tau_r} \exp\left(-\frac{t}{\tau_r}\right) - \frac{2\tau_r}{\tau_1} \exp\left(-\frac{t}{\tau_1}\right)\right]. \tag{7}$$

If $\tau_s > \tau_1$ the system becomes unstable. The behaviour of crystallization rate and interface position under the stepwise change of crystal position is represented in Figure 2.

The transient regimes under the change of the melt temperature are of the most interest. Under the stepwise change δT the changes of crystallization rate

$$\delta v_T = -\frac{\lambda_L}{L h_0} \delta T$$

and interface position

$$\delta H_T = \left[\frac{\lambda_L l_0}{\lambda_s h_0 G_s} - \frac{\lambda_L \tau_L}{L h_0}\right] \delta T$$

takes place in a short time interval τ_L. Then the crystallization rate remains constant during τ_s interval

$$\delta v(t) = \frac{\delta H_T}{\tau_r} \sqrt{\frac{\tau_s}{4t}} \exp\left(-\frac{\tau_s}{4t}\right) + \delta v_T \frac{\sqrt{\tau_s}}{\tau_r} \left[\sqrt{t} \exp\left(-\frac{\tau_s}{4t}\right) - \sqrt{\tau_s} \operatorname{Erfc}\left(\sqrt{\frac{\tau_s}{4t}}\right)\right] \tag{8}$$

until the relaxation processes affect on the crystallization rate and it begins to return to the stationary value v_0. At $\tau_s \ll t \ll \tau_r$ we have

$$\delta v(t) = \frac{\tau_1 + \tau_r}{\tau_1 - \tau_r} \exp\left(-\frac{t}{\tau_r}\right) \left(\delta v_T + \frac{\delta H_T}{\tau_r}\right) - \frac{2\tau_r}{\tau_1 - \tau_r} \exp\left(-\frac{t}{\tau_1}\right) \left(\delta v_T + \frac{\delta H_T}{\tau_1}\right). \tag{9}$$

The behaviour of crystallization rate and interface position under the stepwise decrease of the melt temperature is represented in Figure 3.

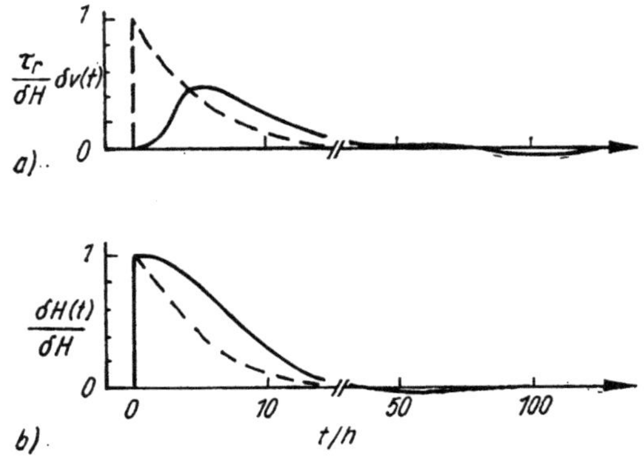

Fig. 2. (see Fig. 1)

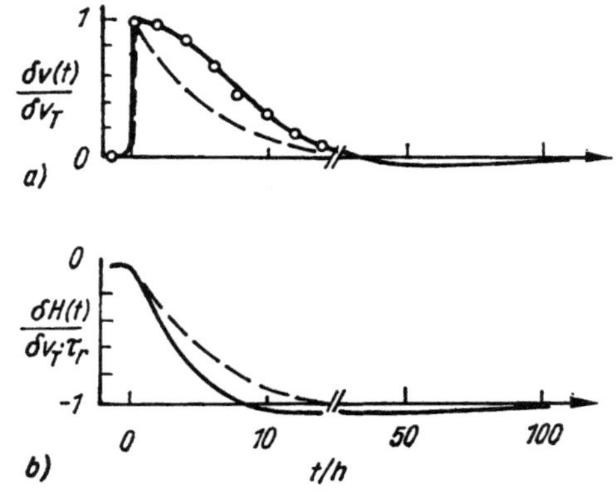

Fig. 3. (see Fig. 1)

Our experimental results, represented in Figs 1a and 3a can be satisfactorily described with expressions (5), (6), (9) and (8) with $\tau_s = 2.2$ h and $\tau_r = 5.4$ h. In our experiments $\tau_s < \tau_r < \tau_1$ and the hesitative nonstability doesn't realize.

The inertia of relaxation process, caused by the large value of crystal diameter, changes the character of crystallization rate and interface position behaviour at transient regimes. The delay of relaxation and described features of transient processes have to be taken into account at programming and operating of growth processes.

References

GORYLETSKY, V., NEMENOV, V., PROTSENKO, V., RADKEVICH, A., EIDELMAN, L.: J. Cryst. Growth **52** (1981) 509

HANGLEITER, T., SPAETH, J.: J. Cryst. Growth **42** (1977) 503

LEYBOVICH, V., POGODIN, A.: Izv. Akad. Nauk SSSR – ser. fiz. **44** (1980) 366

NALBANDYAN, H.: Cryst. Res. Technol. **17** (1982) 815

NALBANDYAN, H.: J. Cryst. Growth **67** (1984) 115

(Received, accepted October 10, 1987)

Authors'' addresses:

Dr. H. G. NALBANDYAN
Condensed Media Physics Institute of Yerevan State University
Shopron str. 21
Yerevan 375090, Armenia, USSR
Dr. V. A. NEMENOV
NPO ''Monokristallreaktiv''
Leninstr. 60
Kharkov 310141, USSR

Cryst. Res. Technol.	23	1988	4	K63–K65

Short Notes

R. K. Marwaha, B. S. Shah

Solid State Physics and Materials Science Laboratories, Physics Department, Saurashtra University, Rajkot, India

Microhardness Studies on Benzoic Acid Single Crystals

Introduction

Molecular solids differ considerably from other types of solids in that the molecule preserves its individuality. Molecular solids comprise the largest group of solid materials because of the almost infinitely large number of organic molecular compounds. The complexity ranges from the monoatomic rare gas solids to the complicated structures of polymeric and biological materials.

The present study is part of an extensive programme to study the mechanical properties of organic molecular solids at low and high temperatures. In this communication are some results on benzoic acid single crystals.

Experiment and results

Single crystals of benzoic acid were grown from the melt by the Bridgman method. The starting material was column chromatographed, twice vacuum sublimed and zone refined. The material was transferred to crystal growth containers without exposure to the atmosphere (Sherwood). The crystals were cleaved with a sharp blade in the usual manner. Smooth (001) cleavages were selected after optical examination. These crystals were then indented on the Carl Zeiss NU2 Universal Research Microscope.

Indents were made with a pyramidal diamond indentor. Loads (P) varied from 5 g to 100 g. A number of indents were made at each load. The mean diagonal length was used in calculating the Vicker's Hardness Number (VHN) using the formula

$$VHN = 1.854 \times P/d^2 \quad (1)$$

where P is the applied load in kilograms and d the mean diagonal length in millimeters. The indentation time of 100 seconds was kept constant and the standard deviation in hardness was calculated at each load.

The specimen size was far larger than the indents made, so as to eliminate the boundary effects on the results. The distance between the indents was kept four to five times the largest size of the indentation mark. The thickness of the specimen was relatively more such that the indentor did not sense the lower surface. A large number of specimen were studied.

Figure 1 is a plot showing the variation of the hardness (VHN) with load (P) for benzoic acid. The plot shows two peaks.

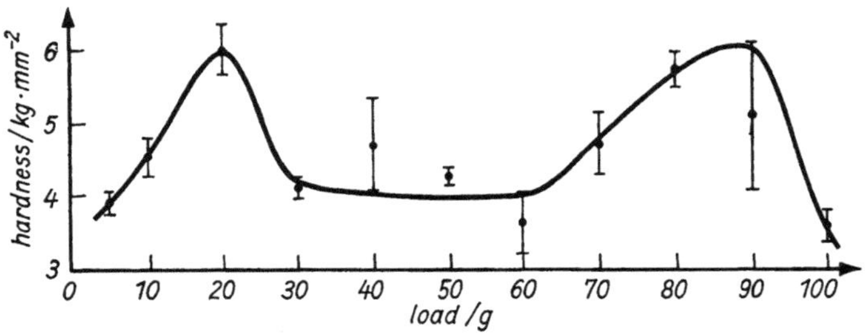

Fig. 1. Plot of Vicker's hardness number versus load

Discussion

Benzoic acid, C_6H_5COOH, belongs to the monoclinic, prismatic class of crystals with space group P 2_1/c and has four monomer molecules per unit cell (SIM et al.). The crystallizing unit is the dimer which consists of a pair of molecules held together by hydrogen bonds. The crystal structure of benzoic acid is shown in figure 2. The dimer unit is similar in size and shape to the p-terphenyl molecule but the arrangement of molecules in the unit cell differs from anthracene, naphthalene, anthraquinone and p-terphenyl. Six-fold coordination is possible in all the planes. The intermolecular forces are weakest in the ⟨001⟩ direction and cleavage occurs readily in the (001) plane.

In producing deformation, only that part of the applied load is effective that can be resolved into shearing forces parallel to the planes in each crystal along which slip is possible. A single crystal just loaded to above its elastic limit, slip occurs first on one or just a few randomly located planes within the crystal. This takes place at low stresses. Once slip begins on a plane, slip continues only for a limited distance and then ceases. As the stress increases, transfer of slip from one potential slip plane to another occurs.

Deformation of benzoic acid has been shown to occur by dislocation movement on three primary slip planes (100), (010) and (001), all of which are crystallographically close packed. The slip systems are (010) ⟨100⟩, (100) ⟨010⟩, and (001) ⟨010⟩. The Burgers vectors' quantities in the ⟨100⟩ and ⟨010⟩ directions are 5.52 Å and 5.14 Å, respectively. The dislocations (100) ⟨010⟩ and (010) ⟨100⟩ are mobile (HANCOCK et al.).

Fig. 2. Crystal structure of benzoic acid

Referring to Figure 1 it is observed that a peak occurs at 20 g load and 90 g load in the plot of the variation of hardness with load. At both these loads the hardness values are nearly the same. The peaks can be accounted for ease of slip at different loads. The peak at 20 g load is due to slip taking place in the $\langle 010 \rangle$ direction for the generation of dislocations of the (100) $\langle 010 \rangle$ type the Burgers vector quantity of which is 5.14 Å. The secon dpeak at 90 g load is due to slip coming into force in the $\langle 100 \rangle$ direction with the generation of dislocations of the (010) $\langle 100 \rangle$ type with Burgers vector quantity of 5.52 Å. The hardness values at the two loads being nearly the same and may be due to the slip on the (100) plane being partly active also.

Conclusions

(a) Two peaks in the hardness versus load plots are observed at 20 g and 90 g load;
(b) The hardness value is nearly the same at both these loads;
(c) The first peak at 20 g load occurs due to slip in the $\langle 010 \rangle$ direction and the second peak at 90 g load due to slip in the $\langle 100 \rangle$ direction;
(d) The nearly same hardness value is due to the $\langle 010 \rangle$ slip being partly still active.

One of the authors (RKM) is grateful to the Principal, Rajkumar College, Rajkot for granting the necessary permission to do research.

References

HANCOCK, P., TEDSTONE, R. D., SHERWOOD, J. N.: Phil. Mag. **23** (1971) 1491
SHERWOOD, J. N.: Purification of Organic and Inorganic Materials, ed. M. ZIEF, New York 1969
SIM, G. A., ROBERTSON, J. M., GOODWIN, T. H.: Acta Crystallogr. 8 (1955) 157

(Received January 7, 1988;
accepted January 14, 1988)

Authors' address:

Prof. B. S. SHAH, R. K. MARWAHA
Department of Physics
Saurashtra University
Kalavad Road
Rajkot 360005, India

Cryst. Res. Technol.	23	1988	4	K66–K69

M. ŁEPEK

Institute of Experimental Physics, University of Gdańsk, Poland

On the Dispersion of the Refractive Index in $A^{II}B^{VI}$ Thin Films

We propose a new calculation following traditional methods for deducing refractive index from the fringe pattern of the transmission spectrum of a thin transparent CdS, ZnS, and $Zn_xCd_{1-x}S$ film surrounded by nonabsorbing media. The error in the refractive index measurements is 4–5%. We have found that polycrystalline CdS, ZnS, and $Zn_xCd_{1-x}S$ evaporated films have refractive index different from those of single crystal material.

Sample preparation

The deposits were generally obtained on glass substrates at temperature 373 K. The experimental set-up consists of an evaporation system (Balzers) capable of attaining vacuum better than 10^{-6} Torr. The source is a cylindrical quartz crucible wrapped in a molybdenum sheet. Details of the procedure and of a systematic study of the layers formed are described elsewhere (DOGIL, ŁEPEK and ŁEPEK, DOGIL). The thickness of film was calculated using the microscopic interference measurement technique. The error in the measurement of the thickness is estimated to be about 5%.

Discussion and results

The optical characterization was performed at room temperature by transmittance measurements in the visible spectrum (0.4–0.8 μm) using UV-VIS double-beam Zeiss spectrophotometer. The spectra were analized in order to obtain the refractive index. Transmission spectra is given in Figure 1. The refraction indexes were determined from the envelope curves T_{min} and T_{max}. We obtained (MANIFACIER et al.):

$$n = \{N + (N^2 - n_0^2 n_1^2)^{1/2}\}^{1/2}$$

with

$$N = \frac{n_0^2 + n_1^2}{2} + 2n_0 n_1 \frac{T_{max} - T_{min}}{T_{max} \cdot T_{min}} . \tag{1}$$

$n_0 = 1$ is refractive index of air and n_1 the refractive index of the glass substrate. Values of n shown in Figures 2, 3, and 4 for CdS, ZnS and $Zn_xCd_{1-x}S$ films respectively. The relative error $\Delta n/n$ for these measurements does not exceed 4–5%.

The value of CdS and ZnS refractive index are higher than the experimental values. The refractive index of a thin films often differs from that of the bulk material. Generally lower values were obtained (PULKER). For example, bulk cubic ZnS has a mean refractive index of $n_{600} = 2.384$ (see DRISCOLL), but in polycrystalline cubic films of 0.5 μm thickness an index of $n_{600} = 2.27$ was determined (MANIFACIER et al.).

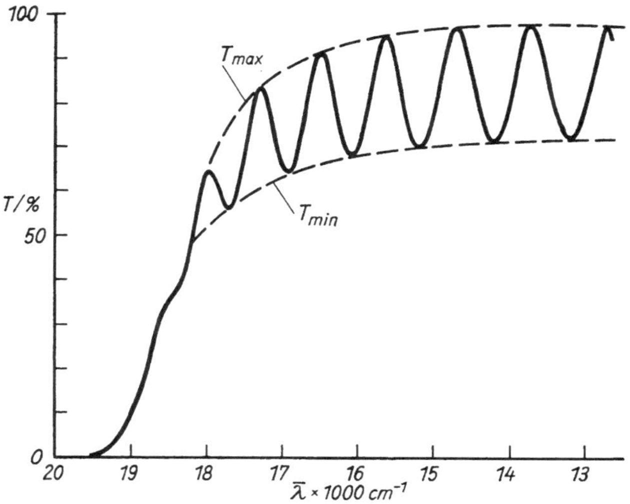

Fig. 1. The optical transmittance of a CdS film of about 1 μm thickness as a function of wavelength

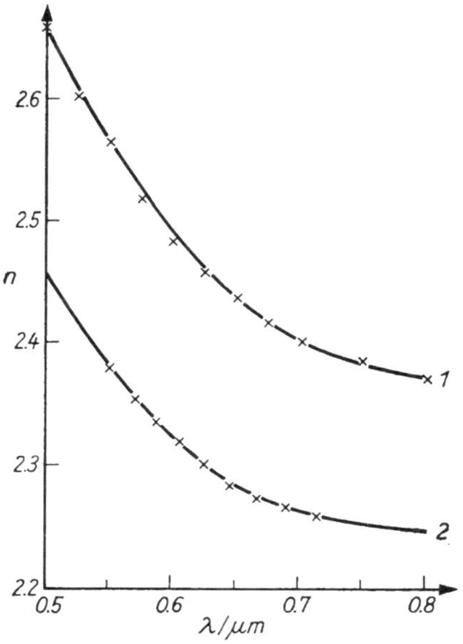

Fig. 2. Dependence of experimental refractive index of a CdS film versus wavelength: curve 1 – values predicted from single-crystal data of CdS (DRISCOLL); curve 2 – values deduced from the interference pattern see Figure 1

The experimental values of the refractive index for the thin films $Zn_xCd_{1-x}S$ shown as a function of the wavelength in Figure 4. For comparison the solid line represents the linear interpolation according to REISINGER. For $Zn_xCd_{1-x}S$ the approximation value by using equation of REISINGER:

$$n^{Zn_xCd_{1-x}S} = xn^{ZnS} + (1 - x)\, n^{CdS} , \tag{2}$$

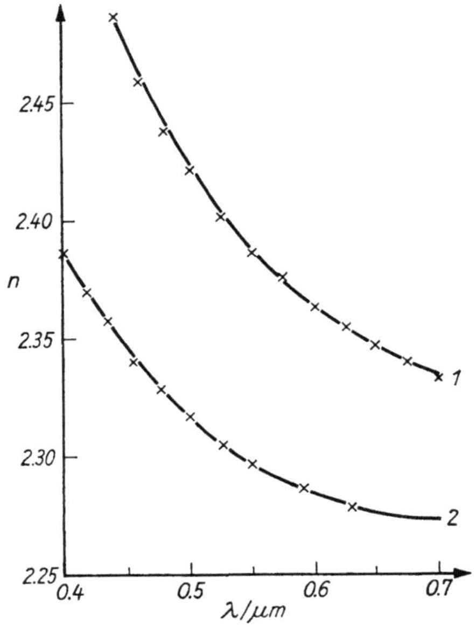

Fig. 3. Dependence of the experimental refractive index of a ZnS film versus wavelength: curve 1 – values predicted from single-crystal data of ZnS (Driscoll); curve 2 – values deduced from the interference pattern

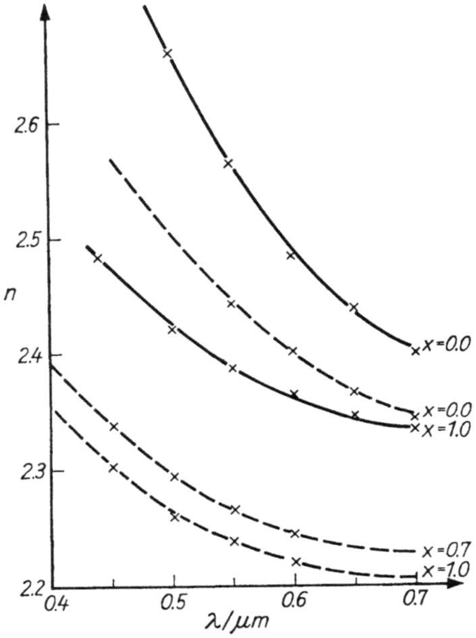

Fig. 4. Dependence of the experimental refractive index of a $Zn_xCd_{1-x}S$ film versus wavelength. The solid lines represent the linear interpolation for $x = 0$ and $x = 1$. The dashed lines specified by $x = 0$, 0.70 and 1.0 indicate the values from equation (1)

taking for n^{ZnS} and n^{CdS} the bulk refractive index of ZnS and CdS respectively, as given by Driscoll. The interpolated refractive index is higher than the measured values. Our experimental values are indicated by dashed line. The obtained values are in agreement with previously published reference data Reisinger. The observed dispersion of refractive index may be explained by probably small deviation from chemical stoichiometry and inhomogeneous structure of the thin films.

Conclusion

The refractive index of a $A^{II}B^{VI}$ thin film often differs from that of the bulk material. All of the experiments on polycrystalline thin films of $A^{II}B^{VI}$ show that the index of refraction depends on the deposition conditions. These are two effects possibly occurring. First, the film may become nonstoichiometric (probably excess of cadmium or zinc). The second property, that varies with the deposition conditions, is the crystalline grain size.

References

Dogil, B., Łepek, M.: Cryst. Res. Technol. **20** (1985) K21
Driscoll, W.: Handbook of Optics, New York 1978
Łepek, M., Dogil, B.: Cryst. Res. Technol. **21** (1986) K127

Manifacier, J. C., Gasiot, J., Fillard, J. P.: J. Phys. E., **9** (1976) 1002
Pulker, H. K.: Appl. Optics **18** (1979) 1969
Reisinger, G. U.: phys. stat. solidi (a) **88** (1985) 745

(Received January 11, 1988;
accepted January 14, 1988)

Author's address:

Dr. M. Łepek
Institute of Experimental Physics
University of Gdańsk
ul. Wita Stwosza 57
80952 Gdańsk, Poland

Cryst. Res. Technol.	32	1988	4	K70–K71

M. ŁEPEK, B. DOGIL

Institute of Experimental Physics, University of Gdańsk, Poland

Structure of CdS Evaporated Films for Ultrasonic Transducers

Vacuum evaporated films of cadmium sulfide have recently been shown to act as efficient high frequency piezoelectric ultrasonic transducer (LEARMANS et al.). To obtain good transducer performance the films must exhibit a high electrical resistivity and a high degree of crystallographic orientation. In the past, several methods of deposition of CdS films by evaporation which give high resistivity films have been suggested. These methods generally consist of multisource coevaporation of Cd and S (BEECHMAN), with or without external doping with silver or copper.

In this paper we report the deposition of high resistivity (10^8 Ω cm) CdS films using a single evaporation technique.

The experimental set-up consists of an evaporation system capable of attaining vacuum better than 10^{-3} Pa. The powder of highly pure CdS is doped with 10 wt% $CdCl_2$ powder. The mixed powders are then pressed to 5 mm pellets by applying a pressure of 3.5×10^7 N m^{-2}. The indium layers are deposited on glass substrate from highly pure indium granules maintained in the tungsten crucible. Next, the CdS film is deposited on indium electrode. The temperature of the quartz crucible is 1223 K, the rate of deposition is maintained at 430 Å s^{-1}. The thickness of the films is about 1 μm. The second indium electrode is deposited on CdS film in the same method as the first one. During each evaporation the temperature of the substrate is maintained at 100 K only. The capacitance of In–CdS–In sandwich was measured in air, its dielectric constant being calculated using the equation for parallel-plane capacitors. All measurements are performed at room temperature.

The samples are investigated using the following three methods: (a) The resistivity measurements are carried out by sandwich In–CdS–In structure with a silver paste having a value of 10^8 Ω cm, which is nearly equal to that reported by Beechman. (b) The dielectric measurements point out that dielectric value of the CdS film is very close to the bulk value and is equal to 10.0 (FORSTER). (c) An X-ray technique using a Debye-Scherrer camera with a fixed beam-to-film plane angle of 22° and with copper filtred radiation are used for studying the structure of CdS films. All CdS films are highly oriented polycrystalline layers with [0001] texture.

Three kinds of observed films are listed in Table 1. X-ray powder diffraction patterns show: (a) the influence of the substrate on the effect of crystallographic orientation, (ii) [0001] texture is visible, (iii) comparison of patterns shows the appearance of cadmium as a second phase in the film deposited on In-layer. For example, a reflex existing at $2\theta = 38.5°$ corresponds to the cadmium index (10.1). The appearance of the second phase can be affected also by heating a sample at 435 K.

In summary, we have established three essential conditions that have to be fulfilled in order to deposit reproducible high resistivity and highly oriented CdS films by this single source evaporation system. These conditions are: (i) The source material has to be rendered excess in cadmium. We accomplished this request by doping

Table 1

No	samples	substrate	thermal treatment	references
a	CdS:Cd	glass	no	this paper
b	CdS:Cd	In-layer on glass	no	this paper
c	CdS	glass	435 K	DOGIL, B., ŁEPEK, M.

CdS powder with $CdCl_2$. (ii) The substrate temperature has to be kept only at 300 K. (iii) The CdS film has to be evaporated on indium layers.

By fulfilling the above three requirements we had been able to deposit reproducible highly oriented thin films of CdS with a resistivity of 10^8 Ω cm.

References

BEECHMAN, D.: Rev. Sci. Instrum. **41** (1970) 1654
FORSTER, N. F.: J. Appl. Physics **38** (1967) 149
LEARMANS, C., MICHIEILS, L., BOCK, A.: Thin Solid Films **15** (1973) 317
DOGIL, B., ŁEPEK, M.: Cryst. Res. Technol. **20** (1985) K21

(Received November 23, 1987;
accepted February 15, 1988)

Authors' address:

Dr. M. ŁEPEK, Dr. B. DOGIL
Institute of Experimental Physics
University of Gdańsk
ul. Wita Stwosza 57
80952 Gdańsk, Poland

Cryst. Res. Technol. | 23 | 1988 | 4 | K72–K73

J. PRZEDMOJSKI, S. GIERLOTKA, B. PURA, R. WIŚNIEWSKI

Institute of Physics, Warsaw Technical University, Warszawa, Poland

High-Pressure Camera for X-ray Investigations of Liquid Crystals

High-pressure X-ray investigations are complementary to other high-pressure experiments on liquid crystals e.g. those performed with optical methods (HERMANN et al.). These investigations, however, are connected with serious exprimental problems. The most important is to obtain satisfactory measurement accuracy from rather low angle diffraction maxima with which we deal in liquid crystals. In which follows we present preliminary results obtained with our high-pressure X-ray liquid crystal cell based on some new concepts concerning the experimental set up.

Standard high-pressure cells, usually adapted for liquid crystal investigations, are equipped with diamond anvils. The deficiency of those cells is that the pressures are higher than required for liquid crystals, and the exposure times are long, about 100 hours (CHANDRASEKHAR et al.; PIEREMARINI, BLOCK). Our camera which is schematically represented in Figure 1 was designed exclusively for liquid crystals.

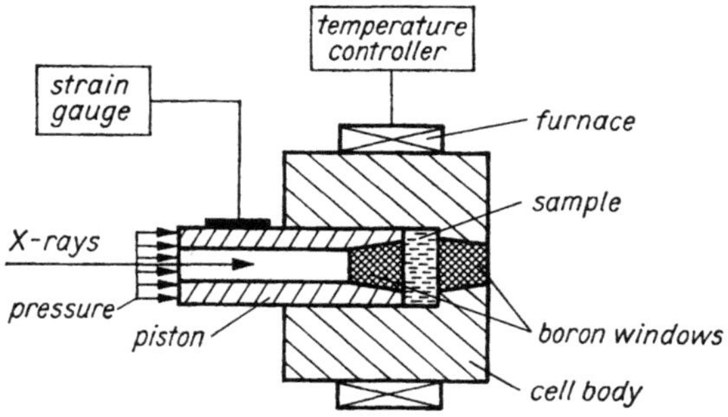

Fig. 1. Scheme of the high-pressure Y-ray liquid crystal cell

The sample is kept in a steel cylinder and the pressure is applied with a piston. Cone chaped X-ray windows are made of a single boron crystal. The cell is heated with the help of an electric furnace. Pressure up to 5 kbar and temperatures up to 120 °C are available. In order to obtain good resolution and short exposure time we applied for the first time in high-pressure X-ray camera Guinier focusing geometry. Using $Co_{K\alpha_1}$ radiation we could obtain good quality patterns like that shown in Figure 2 for a 4 hour exposure.

The camera was tested by using ferroelectric liquid crystal 2M4P8BC with a phase diagram, according to our recent results (GIERLOTKA et al.): Cr–G*–J*–F*–I*–C*–A–Chol–BP–Iso. The effect of pressure on the melting point was investigated. The transformation temperature from solid to smectic G* was deduced from the disappearance of the diffraction lines corresponding to the solid state. The results are

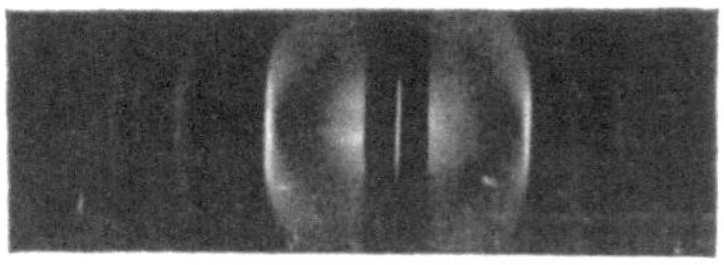

Fig. 2. X-ray pattern obtained in the high-pressure X-ray camera for the solid +2M4P8BC liquid crystal

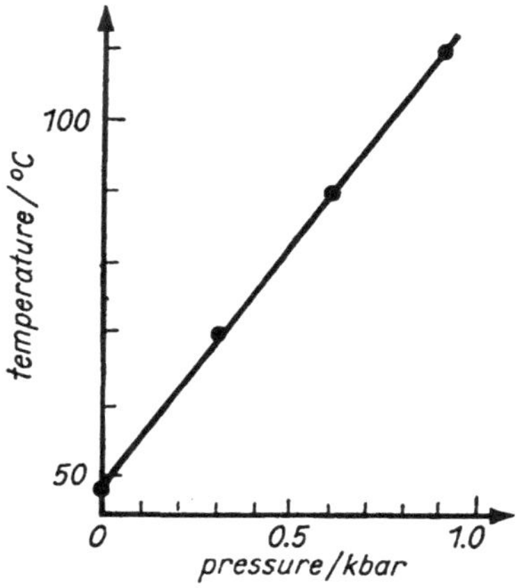

Fig. 3. The plot of the melting temperature versus pressure for +2M4P8BC liquid crystal

given in Figure 3. Although the applied pressures were rather low, the respective changes of melting temperatures were found to be rather large.

References

CHANDRASEKHAR, S., SHASHIDHAR, R., RAO, K. V.: Advances in Liquid Crystal Vol. 1 p. 123, Proceedings of the Third Liquid Crystal Conference of Socialist Countries, Budapest 27–31 August 1979

GIERLOTKA, S. et al.: to be published

HERMANN, J., KLEINHAUS, H. D., SCHNEIDER, G. M.: Journal de Chemie Physique **80** (1983) 111

PIEREMARINI, C. J., BLOCK, S.: Rev. Sci. Instrum. **46** (1975) 973

(Received January 19, 1988;
accepted February 8, 1988)

Author's address:

Dr. J. PRZEDMOJSKI, S. GIERLOTKA, B. PURA, R. WIESNIEWSKI
Warsaw Technical University
Institute of Physics
Koszykowa 75
00-662 Warszawa
Poland

Cryst. Res. Technol.	23	1988	4	K74–K76

Selina Nigli, C. Jagadish, G. K. Chadha

Department of Physics and Astrophysics, University of Delhi, India

Surface Morphology of Melt-grown CdI_2 Single Crystals

Introduction

Optical microscopy is one of the most powerful methods of obtaining direct information on the growth mechanisms of crystals (Komatsu; Sunagawa). In the present paper we have used this method to study the growth mechanism of as-grown and etched (0001) faces of CdI_2 single crystals grown from the melt. In the study of melt-grown crystal surfaces, the disadvantages faced by researchers working with aqueous solution-grown crystals does not arise (Albon, Dunning). The study of the relationship of low step patterns on the crystal faces and the defect structure in crystals has led to the discovery of many new phenomena (Enckevort; Bauser, Strunk). The determination of the nature and distribution of defects is not only necessary for the standardisation of crystal perfection but is equally necessary for a better understanding of growth mechanisms.

The distribution and nature of the defects depends on the nature of the conditions of growth, more often than on the chemical nature of the crystal. For instance, one may distinguish between crystals grown from the melt at high temperatures and crystals grown from solution at temperatures not very different from the room temperature. Many defects are formed during growth. They may be actively involved in the growth process, such as dislocations with a screw component, but are usually a consequence of growth conditions, impurity segregation, inclusions, stacking faults, thermoelastically activated dislocation sources, coalescences of vacancies during cooling or simple handling.

Experimental

The crystals used in the present investigation were grown by the zone-melting technique after suitably zone-refining the material (Nigli; Nigli, Jagadish). CdI_2 being a layered material can be easily cleaved along the basal plane, hence very thin platelets (4–6 mm length and breadth, and 1 → 0 μm thickness) were cleaved and cut for observation by a highly sensitive optical microscope. Microphotographs of as-grown and etched crystal surfaces were recorded. Etching was carried out using standard etching techniques (Amelinckx; Gilman, Johnston). In order to ascertain the single crystal nature of the material X-ray oscillation photographs were recorded using Ni filtered CuKα radiation and a cylindrical camera of radius 3 cm.

Results

The different sections of the crystal ingot examined by X-ray diffraction showed the ingot to be a single crystal with the 4H polytype structure.

Figure 1 is a first example of a growth spiral on melt grown CdI_2 single crystals. Here a group of cooperating spirals start out from an array of dislocations with equal signs. The black dot at the centre is an end-dimple (Frank) of a dislocation meeting the crystal surface. The black deposits are iodine precipitates formed during cooling.

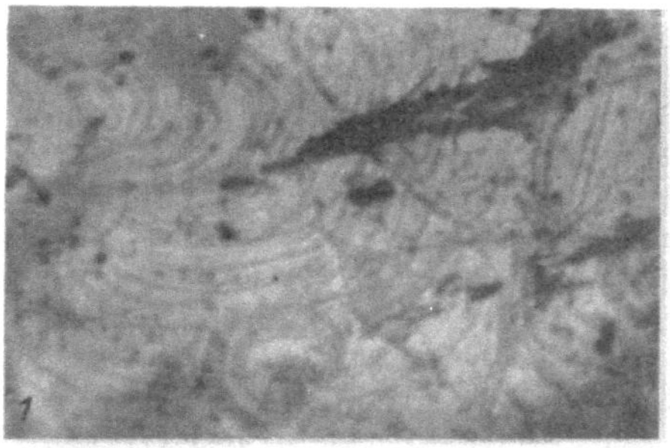

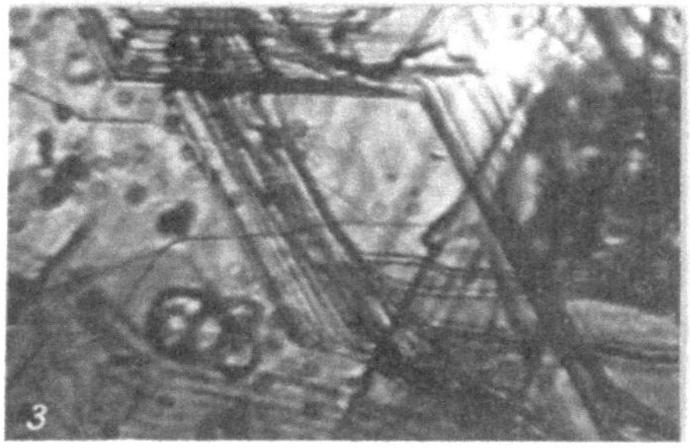

Fig. 1. Microphotograph of as-grown crystal surface showing a partly polygonised spiral

Fig. 2. Detailed view of dislocation centres on as-grown crystal surfaces

Fig. 3. Normal as-grown surface pattern

Fig. 4. Etch patterns of the crystal face

It can be seen that the spiral is partly polygonised. Interlacing of spirals is also seen to take place on one side.

Figure 2 shows a series of spirals on another surface examined. The dislocation centres have been marked by arrows. The normal surface pattern observed in most cases was of the type shown in Figure 3. In most of these micrographs bunching of steps was observed. A large number of etched crystal surfaces were also examined. Some of the patterns obtained are shown in Figure 4.

Discussion

The first growth spiral was observed only after a very large number of crystals were scanned. Hence initially we concluded that melt-grown CdI_2 single crystals were nearly free from dislocations which might act as spiral centres. However, macrosteps were invariably seen. The observation of the first growth spiral, shown in Figure 1, prompted the scanning of many more surfaces which yielded a few more spiral patterns. But when the crystals were etched we found closed hexagonal spirals in all cases. The etch pits are probably located at the centres of the spirals but are not visible as no phase contrast was used. The growth spirals have become visible as a consequence of preferential attack along growth steps.

Macrosteps may arise from high Burgers vectors or from piling up of monolayer steps. Bunching of these steps is caused by the coalescing of dozens or hundreds of steps of unit cell height. The bunching of steps is not completely understood but it is believed to be caused by impurity trapping (Cabrera, Vermileya; Chernov;

VAN EERDEN, MULLER-KRUMBHAAR). CABRERA and VERMILEYA describe step bunching process as that in which impurities adsorb on terraces between monolayer steps. These impurities will retard the motion of the steps. CHERNOV describes the case as that in which the step velocity depends on the step height. He does not specify the physical reason for this. A recent model proposed by VAN DER EERDEN and MULLER-KRUMBHAAR give a definite correlation between bunching and impurity trapping. All the microphotographs are those of the (0001) faces. During crystal growth these faces grow in a direction perpendicular to the direction of movement of the liquid zone. In the direction of the zone there will exist gradients in the impurity level. If the rate of diffusion of the impurities along the direction of movement of the zone is much slower than the rate of growth of the layers, the impurities will be trapped in the growing interfaces. These impurities will act as dislocation centres. The difference in impurity concentration along layers will introduce a time lag in the (screw dislocation) movement of steps along layers. This will be equivalent to a perturbation in the advancing steps leading to the formation of macrosteps. The further advancement of these macrosteps leads to the formation of bunches.

The authors would like to thank Mr. B. Samanta, NPL, for the help with the microphotographs. — Selina Nigli acknowledges with gratitude the suggestions and also the encouragement of Prof. Hermann Neels and Prof. Peter Paufler.

References

ALBON, N., DUNNING, W. J.: Acta Cryst. (1960) 495
AMELINCKX, S.: Acta Met. **2** (1954) 848
BAUSER, E., STRUNK, H.: J. Cryst. Growth **51** (1981) 362
CABRERA, N., VERMILEYA, D. A.: in: Growth and Perfection of Crystals, New York 1958, p. 393–410
CHERNOV, A. A.: Sov. Phys. Usp. **4** (1961) 116
EERDEN, J. P. VAN, MULLER-KRUMBHAAR: Electrochemica Acta **31** (1986a) 1007
EERDEN, J. P. VAN, MULLER-KRUMBHAAR: Phys. Rev. Lett. **57** (1986b) 2431
ENCKEVORT, W. J. P. VAN: Progr. in Cryst. Growth Charact. **9** (1984) 1–50
FRANK, F. C.: Acta Cryst. **4** (1951) 497
GILMAN, J. J., JOHNSTON, W. G.: in: Dislocations and Mechanical Properties of Crystals, (J. C. FISHER et al., eds) New York 1957, p. 116
KOMATSU, H.: in: Crystal Growth and Characterisation, eds R. UEDA, J. B. MULLIN, Amsterdam 1975, p. 333
NIGLI, S.: Ph. D. thesis, 1986
NIGLI, S., JAGADISH, C., CHADHA, G. K.: Cryst. Res. Technol. **22** (1987) K219
SUNAGAWA, I.: in: Crystal Growth and Characterisation, eds UEDA, R., MULLIN, J. B., Amsterdam 1975, p. 347

(Received February 10, 1988;
accepted February 15, 1988)

Authors' address:

Dr. G. K. CHADHA*), Dr. SELINA NIGLI, G. JAGADISH**)
Dept of Physics & Astrophysics
University of Delhi
Delhi-110007
India

*) *To whom all communications may be sent:*
**) *Present address*:
Department of Physics and Astrophysics
Sri Venkateswara College
Dhaula Kuan
New Delhi 110021, India

Cryst. Res. Technol. | 23 | 1988 | 4 | K77–K79

Letter to the Editor

Unexampled Plagiarism in the Theory of Electromagnetic Waves

In 1983 in the USA the book "Theory of Electromagnetic Waves. A Coordinate-Free Approach" (Mc Graw Hill Book Company, 449 pp.) by Hollis C. CHEN, professor on the Ohio University (provost J. L. Bruning) was published. It contains 11 chapters. Chapter 1. "Linear Analysis" includes the description of mathematical apparatus of covariant coordinate-free methods of tensor calculus. In small Chapter 2. "Basic Equations of Electrodynamics" a well-known information about Maxwell equations is given. In the next four Chapters: 3. "Some General Properties of Plane Waves", 4. "Plane Waves in Isotropic Media", 5. "Plane Waves in Crystals" and 6. "Plane Waves in Uniaxial Media" the propagation of electromagnetic waves in isotropic and anisotropic media is considered on the base of the methods presented in Chapter 1. In the last five chapters the obtained results are applied to some specific cases of propagation of electromagnetic waves in plasma, ferrites, moving medium (Chs. 7, 8) and to the radiation theory in isotropic, uniaxial and moving media (Chs. 9–11). The bibliography contains about 300 points.

As follows from this list, the major part of the book contents concerns the investigation of propagation of electromagnetic waves in anisotropic media, i.e. in crystals, and therefore is closely connected with the subjects of journal "Crystal Research and Technology".

At the jacket of the book its brief abstract is given where it is stated that "this new book, for the first time, treats the subject matter completely from a coordinate-free approach". This quotation leaves any doubts that professor H. C. CHEN ascribes to himself the merit of creation and development of a new effective coordinate-free approach in such important region of science and technology as theory of propagation of electromagnetic waves in a media with different kinds of anisotropy.

However, it is the covariant coordinate-free approach to the theory of electromagnetic waves in a media with arbitrary anisotropy that was for the first time proposed, thoroughly grounded and developed by Soviet scientist F. I. FEDOROV more than 30 years ago. His first publication concerning the subject has appeared in 1952 [1]. In 1954 he has defended Doctor Thesis "Invariant methods in optics of anisotropic Media" [2]. Later, as a result of development and improvement of his covariant method, F. FEDOROV has published about 100 papers and 5 monographs [3–7], first of which was published 25 years earlier than Chen's book.

FEDOROV's works are well-known in the USSR and other countries. Proposed and developed by him the method was named "a covariant method by FEDOROV". He is the acknowledged head of a scientific school that developed this lead created by him, the representatives of this school work in the largest scientific centres of the USSR. Nevertheless the bibliography in Chen's book contains no references to either of FEDOROV's works. How can be explained such fact ? Didn't Chen know this well-known theory that treats the propagation of electromagnetic waves, i.e. the problem, which his own book is devoted to ? This assumption falls away at once after the confrontation of CHEN's book with the books and papers by F. FEDOROV. Coincidence and direct adoption from FEDOROV's works begin in the first lines of CHEN's book. Chen's Preface begins with the sentences that are word by word translation into English of the corresponding sentences from Introduction to [3] or their minimum modification that doesn't change absolutely the sense and is used only for masking. Analogously, in text at the jacket the novelty, effectiveness and advantages of covariant approach are characterized essentially in the same expressions as in [3, 4] where they are borrowed from, but they are ascribed to CHEN's merits.

Contents of CHEN's first chapter is completely recopied from §§ 25, 27 of [4]. Variations made by CHEN in the copied text are reduced to using of other designations for vector operations and of another system of units. Thus instead of ab, $[ab]$, $a \cdot b$, n^* Chen uses $a \cdot b$, $a \times b$, ab, $n \times]$, respectively and so on. Furthermore, some letters are changed. Besides, in some cases he makes a rearrangement or regrouping of formulae and corresponding text. Finally, additive formulae are sometimes inserted that are the direct and trivial consequences of FEDOROV's ones. In this part H. CHEN has recopied from [4] about 200 formulae. Analogously, text by CHEN, that accompanies the formulae represents slightly modified translation into English of respective explanations from [4]. Thus professor H. CHEN has impudently ascribed to himself the covariant approach to the theory of electromagnetic waves, that F. FEDOROV had proposed, developed and perfected in the result of many years work.

Just the same is referred to Chapters 3—6 of CHEN's monography. Their contents are completely recopied from the first four chapters of [3]. Totally more than 800 formulae and corresponding text are stolen here.

As a summary it should be noted that more than one half of CHEN's book, it's main scientific concept, more than 1000 formulae and corresponding text are the plagiarism from FEDOROV's monographs [3] and [4].

The appearance of such unprecedented plagiarism has induced me to become acquainted with the other publications of H. CHEN, and it has become clear that as far as in 1980 he had made a report at the U.R.S.I. International Symposium in Munich [8]. This report was republished in 1981 in the journal "Radio Science" (Editor A. T. WATERMAN). The whole contents of the report and the paper, all the 22 formulae and the only figure are recopied from §§ 17, 6, 30 of [3]. Then in 1982 H. CHEN has published in the "Journal of Applied Physics" (editor L. GUTTMAN) a paper [9] all the 38 formulae of which were recopied from §§ 35, 36, 16, 39 of the same book [3]. Finally, in 1983 in "IEEE Transactions of Microwave Theory and Techniques" (editor T. ITOH) one more paper by CHEN was published [10], in this case it was recopied from §§ 17, 21, 6, 26 of just the same FEDOROV's book. However, these results were obtained and published by F. FEDOROV as far as in 1952 [1], and then they were included in the book [3]. Thus in this case H. CHEN has set up a peculiar record by representing as his own the original results obtained and published by the well-known scientist in an internationally known journal 31 years earlier! It should be added that there are no references to FEDOROV's works in all these publications. It's hardly to believe that this can take place in the XXth century!

The author must confess that neither he nor his colleagues know in the history of science an example of plagiarism of such incredible scale. Indeed, in this case the question is that the whole scientific theory developed by F. FEDOROV on the base of an original method proposed by him as a result of many years work, and published in a number of monographs and many papers, has been stolen by H. C. CHEN in sight of the whole scientific world.

We are sure that world scientific community and, first of all, our respected American colleagues will give proper estimation to this unprecedented plagiarism.

References

[1] FEDOROV, F. I.: Definition of Optical Parameters of Uniaxial Crystals by Reflected Light, Doklady AN SSSR, v. 84, No. 6, pp. 1171—1174, 1952

[2] FEDOROV, F. I.: Invariant Methods in Optics of Anisotropic Media, Doctor Thesis, Leningrad 1954
[3] FEDOROV, F. I.: Optics of Anisotropic Media, Izd. AN BSSR, Minsk, 1958, – 380 p.
[4] FEDOROV, F. I.: Theory of Gyrotropy, Nauka i Technika, Minsk, 1976, – 456 p.
[5] FEDOROV, F. I., FILIPPOV, V. V.: Reflection and Refraction of Light by Transparent Crystals, Nauka i Technika, Minsk, 1976, – 222 p.
[6] FEDOROV, F. I.: Theory of Elastic Waves in Crystals, Nauka, Moskva 1965, – 386 p.
[7] FEDOROV, F. I.: Theory of Elastic Waves in Crystals, Plenum Press, New York 1968, – 375 p.
[8] CHEN, H. C.: A Coordinate-Free Aproach to Wave Reflection from an Anisotropic Medium, Int. URSI Symp. Munich, 1980, El.-mag. Waves, 142C/1–142C/3; Radio Science, v. 16, pp. 1213–1215, 1981
[9] CHEN, H. C.: A coordinate-free approach to wave propagation in anisotropic media, J. Appl. Physics **53** (7), July, 1982
[10] CHEN, H. C.: A Coordinate-Free Approach to Wave Reflection from a Uniaxially Anisotropic Medium, IEEE Trans. Microwave Theory and Techniques, v. MTT-31, No. 4, April 1983

I. S. REZ, Moscow

Crystal Research and Technology

eitschrift für experimentelle und technische Kristallographie

Edited by H. NEELS, Leipzig

Papers in Number 5, 1988

Original Papers

AKADEMIE-VERLAG · BERLIN

Continued from cover page 4

Short Notes

Letter to the Editor

Book Reviews

Crystal Research and Technology
Volume 23 1988 Number 4

Contents

Original Papers

Continued on cover page 3

Crystal Research and Technology is indexed in Current Contents/PC & ES.

Bei Fragen zur Produktsicherheit wenden Sie sich bitte an:
If you have any questions regarding product safety,
please contact:

Walter de Gruyter GmbH
Genthiner Straße 13
10785 Berlin
productsafety@degruyterbrill.com